EVOLVING

D1115994

EVOLVING

THE HUMAN EFFECT
AND WHY IT MATTERS

DANIEL J. FAIRBANKS

Prometheus Books

59 John Glenn Drive
Amherst, New York 14228–2119

Published 2012 by Prometheus Books

Cover images © 2012 Media Bakery and Wood River Gallery
Cover design by Grace M. Conti-Zilsberger

Inquiries should be addressed to
Prometheus Books
59 John Glenn Drive
Amherst, New York 14228–2119
VOICE: 716–691–0133
FAX: 716–691–0137
WWW.PROMETHEUSBOOKS.COM

16 15 14 13 12   5 4 3 2 1

Library of Congress Cataloging-in-Publication Data

Fairbanks, Daniel J.
    Evolving : the human effect and why it matters / by Daniel J. Fairbanks.
       p. cm.
    Includes bibliographical references and index.
    ISBN 978–1–61614–565–1 (paper : acid-free paper)
    ISBN 978–1–61614–566–8 (ebook)
    1. Evolution (Biology)—Philosophy. 2. Evolution (Biology) 3. Human evolution.
4. Evolution (Biology)—Social aspects. 5. Human genome. 6. Health. 7. Food supply.
8. Nature—Effect of human beings on. I. Title.

QH360.5.F24 2012
576.8'2—dc23
                                                                      2012000378

Printed in the United States of America on acid-free paper

CONTENTS

PREFACE

Not long ago, I was part of a group of evolutionary biologists that presented an afternoon workshop at a conference for high school science teachers. The teachers were from various places in the northeastern United States—nearly all were from public high schools in locations ranging from the largest East Coast cities to rural towns. Etched in my mind are the words and faces of those teachers, their expressions portraying the personal side of a phenomenon all too familiar to me. They felt weary and embattled, under incessant pressure for having the courage to teach biology the way it should be taught—in the framework of evolution.

Every so often, I read an article or hear a program about the search for a unifying theory in physics, a single explanation for all physical phenomena, from the most miniscule subatomic particles to the unimaginable vastness of the intergalactic universe. While physicists may still be searching for a unifying theory, biologists are not. We have it, and it is evolution, the common thread that ties together all aspects of biology.

Yet the teachers I mentioned a moment ago cannot teach our discipline's unifying theory without enduring ongoing criticism, even threats, from students, parents, fellow teachers in other disciplines, school administrators, school boards, political candidates, elected and appointed government officials, legislators (both state and national), and church ministers. And the teachers at this conference were from the northeastern United States, where acceptance of evolution is higher than in most other regions of the country. Opponents of evolution are constantly at work to invent so-called alternative theories and to craft legislation that, if enacted and enforced, would require science teachers to teach unscientific notions as if they were legitimate. As I write these words, antievolution education bills are pending in the state legislatures of Indiana, New Hampshire, Missouri, and Oklahoma, while a signifi-

7

cant number of aspiring politicians, ranging from school board to presidential candidates, promote opposition to evolution as an integral component of their campaigns. Ironically, this opposition happens at a time when the already abundant and powerful evidence of evolution—especially human evolution—is growing at a rate far greater than at any other time in the history of science.

This fast-paced accumulation of evidence has left the general public largely unaware of how extensive and deep the evidence is. Much of the evidence lies in our DNA and in the DNA of other species—and is revealed through the massive sequencing efforts of scientific teams collaborating on genome projects. While the quantity of DNA sequencing information has exponentially increased in recent years, paleoanthropologists have been amassing extraordinary fossil evidence. Newly discovered fossils of previously unknown extinct humanlike species have been described in the past ten years. And these two mostly independent lines of evidence—DNA and fossils—have recently converged with the sequencing of the Neanderthal genome.

There are two principal reasons why so many people are unaware of the overwhelming evidence supporting evolution. Either they consciously choose to reject it without examining it, or they have difficulty understanding it. This book is written for those in the latter category, who (like most students in my general biology courses) claim that "science is not my thing." My intent is to take admittedly complex scientific information and accurately present it in a way that makes sense to a casual reader. Moreover, the final chapters of this book explore why an understanding of evolution matters, and why efforts to oppose it should concern us all.

Those who deserve the highest recognition for the information in this book are the scientists who discovered it. Many of them travel the world to excavate fossils or collect DNA, and they work long hours in the laboratory and with sophisticated computer applications to generate and analyze reliable data. I am proud to be one of them.

I wish to thank the editors and staff of Prometheus Books for their professionalism, efficiency, and friendship in all aspects of publishing. Dr. Duane Jeffery merits special thanks for his assistance with the manuscript, as does Dr. Donna Fairbanks as a supportive and fearlessly critical nonscientific reader. They have been especially helpful in pointing out errors, omissions, and difficult to understand passages. For any errors that remain, I take full responsibility. The opinions in this book are mine, and, although they are consistent with the opinions of scientists in general, they do not necessarily represent those of the publisher or my employer.

PROLOGUE

Alittle over a century and half ago, Charles Darwin published the first edition of *The Origin of Species*, which would become one of the most influential books in history. He rushed the work to print ahead of his intended schedule because Alfred Russel Wallace had arrived at the same idea—that species were not fixed but instead had arisen by evolution through natural selection—and was about to ruin Darwin's claim to originality. Wallace had sent Darwin a twenty-page manuscript, hoping he would like it and forward it to Charles Lyell for publication. Wallace's words, which Darwin read that June day in 1858, struck him like a paralyzing bolt of lightning. In his letter to Lyell recommending the manuscript's publication, Darwin lamented, "I never saw a more striking coincidence: if Wallace had my MS. [manuscript] sketch written out in 1842, he could not have made a better short abstract! Even his terms now stand as heads of my chapters. . . . So all my originality, whatever it may amount to, will be smashed."[1]

Neither Darwin nor Wallace treated each other as adversaries. A little more than a year earlier, Darwin had written to Wallace, "I can plainly see that we have thought much alike & to a certain extent have come to similar conclusions. . . . This summer will make the 20th year (!) since I opened my first notebook, on the question how and in what way do species and varieties differ from each other—I am now preparing my work for publication, but I find the subject so very large."[2] Wallace replied with a question that hit straight to the core of a fear Darwin had held for some time. Would the book include human evolution? Although Darwin freely admitted his conclusion that evolution applied to all of life, humans included, his response reveals unmistakable hesitation: "I think I shall avoid the whole subject."[3] Darwin was almost true to his word. He entirely skirted the subject in *The Origin of Species*, except for one tantalizing sentence: "Light will be thrown on the origin of man and his history."[4]

When Darwin wrote these words, he could hardly have envisioned how much light would be thrown on our origins over the next century and a half. Use of the word *overwhelming* to describe the accumulated evidence of human evolution borders on understatement. In Darwin's day, the evidence of human evolution consisted mostly of the anatomical and physiological similarities we share with other mammals, especially other primates, and a few scattered humanlike fossils yet to be properly classified as ancient relatives of humans. Today this evidence has grown tremendously to become abundant and powerful, including data from fossils, archaeological excavations, geography, detailed anatomical studies, biochemistry, radiometric dating, cell biology, chromosomes, and massive amounts of information from DNA, especially from genome projects. For those familiar with the evidence, there is no doubt whatsoever—*we evolved.*

But does it matter? Couldn't we just leave today's highly technical evidence of human evolution to scientists and get on with more important issues? To put it bluntly—*it has never mattered more.* Studies of how our genome evolved are revealing why we get cancer and how we may be able to prevent, treat, and cure it. Our evolution is interwoven with the evolution of other organisms, especially the microorganisms that infect us and cause serious diseases. Evolutionary theory predicts that these organisms will persistently evolve to overcome our efforts to control them—and this is now happening. Some bacteria have recently evolved to resist even our most powerful antibiotics. The HIV virus, which causes AIDS, evolved to infect humans less than a century ago, and it continues to evolve as we try to combat it. Other viruses are following suit—and the evidence shows that viruses have been evolving for millions of years in the same manner as they do now. An understanding of our own evolution, and how it is connected to the disease-causing agents that infect us, offers the promise for finding strategies to counter them.

And it's not just us. Our ancestors intentionally shaped the evolution of plants and animals we consume as food—and modern scientists have hastened the evolution of these plants and animals. A plethora of scientific information tells us where, how, and when each plant or animal we use for food evolved from wild into domesticated species. Moreover, each of these species has its own set of disease agents that infect it. As plant and animal breeders seek genetic resistance to diseases in the species they breed, infectious organisms evolve to overcome the resistance. Protecting our food supply against this evolutionary onslaught is a never-ending battle—and an understanding of evolution helps us win it year after year, thanks to persistent scientists. Areas of the earth that once were wet are drying up. In other places the reverse is under way. We must find ways to adapt, to better manage our water and energy resources, to breed crops that are more tolerant of drought, heat, salty soils, and other factors that threaten them. Again, what we have learned from evolution helps us meet this challenge.

In the midst of technological globalization and climate change, we see species threatened with extinction, and actually going extinct, at an alarming rate. The fossil record tells us this is not the first time in the earth's history that species have suffered mass extinction, nor is it the first time our human ancestors have faced climate change. We also learn from the fossil record that we have much to fear if we do not act to deal with these issues. And an understanding of evolution gives us the tools to best conserve the resources of our planet and the diversity of life that so enriches it.

Along with the rest of life, we evolved and we are evolving. And it matters—our future depends on how we use our understanding of evolution.

CHAPTER 1

WHAT IS EVOLUTION?

As a child, I lived in paradise—acres of untouched nature surrounding our home. The clear, cold water of a natural artesian well gushed through a bed of sand to feed streams that cascaded down a hillside into a pond nestled among the trees. During the long summer days when most of my friends were playing Little League soccer, I spent hours hiding behind clumps of tall grass near the pond, watching wild animals emerge from their hiding places to carry on with their lives.

Having been raised in a family of artists, I often toted a sketch pad and pastels with me and drew detailed pictures of the plants and animals I observed. I made leaf presses; collected insects; and cared for our vegetable garden, grapevines, and fruit trees. I raised dogs, pigeons, parakeets, canaries, rabbits, turtles, fish, snakes, lizards, gophers, and monarch butterflies. I caught the butterflies as caterpillars on milkweed plants near our house and watched day after day as each one developed into a gold-spotted, pale-green chrysalis, then into an orange, black, and white adult. My heart filled with magnificent childhood wonder when I released the adult butterflies and watched them disappear into the wind. By the time I was in grade school, my fate was sealed—I was destined to be a biologist.

I sensed a kindred spirit when I learned of the Russian biologist Theodosius Dobzhansky. He collected butterflies, moths, and beetles during his childhood and read Darwin's *Origin of Species* at the age of fifteen. In the turbulent aftermath of the Russian Revolution, he left the Soviet Union with his wife and daughter to study evolutionary genetics in the United States and remained there for the rest of his life. He rose to become one of the most renowned scientists in history, a prolific author of articles and books, several of which are now

15

science classics. In 1973, two years before his death, he penned a short article whose title itself has become a classic—"Nothing in Biology Makes Sense Except in the Light of Evolution."[1] Nearly all biologists concur—evolution is the central theme that unifies biology and explains the whole of life.

What is evolution? Like many broad terms, the word *evolution* can be defined at several levels. At its simplest level, it means genetic change over generations. However, most people perceive evolution—correctly—as the logical extension of this simple definition: *Genetic change over many generations ultimately results in the emergence of new and different species from a single ancestral species.* Not all species become ancestors of others, however. The ultimate fate of most species is extinction.

The evidence of evolution we can observe within a few generations—such as mutations in DNA, shuffling of genes, the interaction of organisms with one another and their environments, and the unequal survival and reproduction of individuals—are all parts of evolution that scientists can readily test and observe. If these processes result in the eventual emergence of new species, we ought to find evidence of common ancestry in the fossil record, in the anatomical characteristics of similar species, and in the genetic information carried in DNA of modern species. By amassing, analyzing, and connecting these various lines of evidence, scientists can reconstruct much of the evolutionary history of life.

For nonscientific reasons, some people belittle evolution as "just a theory." Yet to scientists the word *theory* has a very powerful meaning: *A theory is a well-supported conceptual framework that encompasses a large body of scientific facts, inferences, observations, and experiments and explains them in a coherent way.* Scientifically, evolution is both a fact and a theory. It is a fact in the sense that many of its various components have been repeatedly observed and are so well documented that they can no longer be reasonably disputed. It is a theory in that the various facts, laws, and tested hypotheses bearing on evolution can be broadly explained as manifestations of the same overall process. The same can be said for other broad scientific theories. Substitute the words *gravity* or *the atomic nature of matter* for *evolution* in the preceding sentences of this paragraph, and the sentences become valid scientific expressions regarding these theories.

How did the theory of evolution rise to such prominence that it is now the central theme of biology? The name most often associated with evolution is Charles Darwin. As mentioned in the prologue, he published in 1859 *On the Origin of Species by Means of Natural Selection, or the Preservation of Favoured Races in the Struggle for Life*, which we now refer to simply as *The Origin of Species*. Although this book was the most powerful exposition of evolution in its day, the idea of evolution did not originate with it. There are some hints of evolution in the writings of Greek philosophers, but emergence of modern evolutionary ideas dates to the eighteenth century. Prior to this time, the prevailing view of life was known as *special creation*, aptly described by the seventeenth-century English philosopher John Ray as "the Works created by God at the first, and conserv'd to this day in the Same State and Condition in which they were first made."[2] Most people believed that the earth was no more than a few thousand years old, and all organisms, including humans, are essentially the same now as they were when the earth originated.

Those who studied nature during the nineteenth century were called naturalists; the term *biologist* was not yet in widespread use. Noticing the obvious hierarchical nature of life, several naturalists attempted to categorize living species into increasingly larger groups that made sense based on physical similarities. For example, different species of butterflies are more similar to one another than they are to moths, an observation justifying the classification of all butterflies into one group and all moths into another. Butterflies and moths are more similar to each other than either is to houseflies. And houseflies are more similar to fruit flies (both have only two wings) than either is to wasps and bees (which, like most winged insects, have four wings). At a higher level, all are insects and have the common anatomical characteristics that define insects. Arachnids (such as spiders, mites, and ticks) and crustaceans (such as shrimp, lobsters, and crabs) are not insects, but they have many characteristics in common with insects, such as compound eyes and a segmented exoskeleton made predominantly of a substance called chitin.

Naturalists united the insects, arachnids, and crustaceans into a larger group called the arthropods. They identified other hierarchical groups among plants and animals. Humans, for example, are included within the primates (humans, apes, monkeys, galagos, and lemurs), and the primates within the mammals, and the mammals within the vertebrates. Recognizing that offspring resemble parents, and that close relatives are more similar than distant relatives, naturalists began to raise a question. Might the obvious hierarchy among species throughout all of life be evidence of ancient hereditary relationships? Could different species with similar characteristics have shared common ancestry? If so, might all living things be related to one another in a great tree of life?

A key to understanding evolution is the *biological species concept*, which defines a *species* as a group of individuals that can mate and produce fertile offspring. This is the definition that guided naturalists in the eighteenth and nineteenth centuries and that to some extent is still valid. However, it is a definition crafted by scientists to explain natural phenomena, and nature does not always follow the rules we use in our attempts to explain it. The lines between species are often blurred. Some closely related species may mate and produce infertile offspring. Perhaps the best known are horse-donkey matings, which produce mostly infertile mules.

Mules are but one of many such instances in nature. In other cases, matings between two related species produce offspring with partial or reduced fertility but not a complete loss of fertility. For example, a mating between a female tiger and a male lion produces a hybrid liger. Female ligers retain some fertility and can sometimes mate with male lions or male tigers and produce offspring. By contrast, male ligers are infertile, so ligers cannot mate with each other and produce offspring. Even today, scientists find it difficult to identify at which point many closely related groups of animals or plants have separated sufficiently to define them as separate species.

Naturalists in the eighteenth and nineteenth centuries were well aware that animal breeders had developed self-propagating breeds of animals. Among dogs, for instance, if purebred female and male golden retrievers mate, their offspring are all purebred golden retrievers. The same is true for pure breeds of cats, cattle, sheep, horses, and other domesticated animals, as it is for domesticated plants. Hybridization between different breeds of the same species can likewise produce fully fertile *hybrid* offspring. When I was a child, I had a dog who was a cross between a German shepherd and a dachshund. A man (who knew little about genetics) had intentionally allowed a female German shepherd to mate with a male dachshund, hoping to see miniature

German shepherds in the offspring. Instead, my dog was a medium-sized, floppy-eared, short-legged animal with thick tan hair—a true intermediate between his two parents. We named him Tiegel (German for *skillet*), and I loved him as only a child could. Had he mated with a female German shepherd/dachshund hybrid like himself, their offspring would not have bred true. Instead, a wide array of types would have resulted in the next generation.

Observing the ability of pure breeds to breed true, and the inability of hybrids to do so, some naturalists began to wonder. Could similar species simply be well-advanced breeds (or *varieties*, the term Darwin correctly used to refer to both animals and plants), separated from one another for so long that they have lost the ability to produce fertile offspring when hybridized? Clues that this might be the case were the blurred lines between some species, such as horses and donkeys or lions and tigers. And if new species could diverge from a common ancestral species, how far could the divergence extend? Could all of life ultimately have a common, ancient origin?

Also during the eighteenth and nineteenth centuries, geologists were actively exploring the earth for fossils. Most fossils resembled modern organisms but clearly were not the same. Dinosaurs, for example, were obviously vertebrates and had many characteristics of reptiles and birds, but no living dinosaurs had been found. The same was true for nearly all other fossils— although the species was no longer present, the fossils resembled living species in many ways. Moreover, fossils were found almost exclusively in sedimentary rocks: layered rocks that formed by deposition of sediments. In the early 1800s, several geologists, among them George Cuvier and Alexandre Brongniart in France, and William Smith in England, documented the progression of different fossils in sedimentary layers and concluded that these layers were deposited over long periods of time. They further concluded that most fossilized species had gone extinct and were replaced by a succession of new species.

In 1830, Charles Lyell published the first volume of a massive three-volume work on geology offering evidence that the earth must be at least millions of years old. With a very old earth, might there have been sufficient time for different types of animals and plants to emerge from a single species and eventually diverge into new species? If so, this process would explain both the hierarchical nature of life today and the similarities of modern life to the fossilized remains of ancient life.

Among the early evolutionists were Jean-Baptiste Lamarck and Erasmus Darwin (Charles Darwin's grandfather). Unfortunately, Lamarck is often ridiculed today, perhaps unjustly. He was a brilliant naturalist who developed one of the first coherent hypotheses to explain evolution. With little understanding of the mechanisms of inheritance (which were not well documented in his day), he proposed that animals evolve through the inheritance of characteristics acquired through the use and disuse of organs. For example, he hypothesized that because sightless moles have useless remnants of eyes, the disuse of eyes in their ancestors caused them to lose their eyesight, and their offspring then inherited that trait. According to modern evolutionary theory, Lamarck was correct in his assertion that the distant ancestors of moles had functional eyes, and the sightless eyes in modern moles evolved as remnants of that time. However, the difference between modern evolutionary theory and Lamarck's idea is the *cause* for the loss of eyesight. According to Lamarck's thinking, the light-free environment directly *caused* moles to lose their eyesight, and this environmentally induced change was thenceforth inherited. According to modern evolutionary theory, undirected mutations in DNA, rather than environmentally instigated changes in the inherited material, caused the loss of eyesight. Because there was no advantage for functional eyes in a light-free environment, these inherited mutations persisted, but the environment did not directly cause them.

Charles Darwin was born on the same day as Abraham Lincoln, February 12, 1809, to a prestigious and wealthy family in England. His father was a well-known physician who hoped that his son would pursue the same profession. After a year of medical studies, however, Charles became disillusioned and abandoned them. For a time, he studied to be an Anglican priest but soon left that pursuit as well. His interest in naturalism had already begun during this time as he took courses on natural history and collected beetles. Charles never knew his famous grandfather, Erasmus Darwin, who died seven years before Charles was born, but his grandfather's writings had a strong influence on him. In 1831, as a youth in his twenties, Charles accepted the invitation to serve as naturalist on a round-the-world voyage aboard the HMS *Beagle*, a ship whose crew was assigned to explore the coastline of South America. Darwin received a copy of the first volume of Lyell's work on geology, which had just been published, and studied it extensively on the voyage.

While on the nearly five-year journey, Darwin kept copious notes on the plants, animals, and fossils he saw and made large collections of specimens, which he regularly sent home to England at the ship's various ports of call. At the time, he had not fully formulated his ideas about evolution, but his observations on the trip tremendously impacted his thinking. Especially influential were the Galápagos Islands, a group of volcanic islands in the Pacific Ocean, west of the coast of Ecuador. By the time Darwin reached the Galápagos, he had explored most of the South American coastline and was very familiar with the animals and plants native to the continent. On the Galápagos, he noticed a pattern among the animals. Those on the islands strongly resembled those on the mainland, but the island types were much more diverse even though the land area of the islands was much smaller. For example, Darwin collected what were later determined to be thirteen different species of finches on the Galápagos, compared to a relatively few species on the mainland, and he noted that "a nearly perfect gradation of structure in this one group can be traced in the form of the beak, from one exceeding in dimensions that of the largest gros-beak, to another differing but little from that of a warbler."[3]

At the time, Darwin did not venture to explain the similarity of the island types to those on the mainland or why there was so much variation on the islands. But after his return to England, he shared his notes and specimens with other naturalists and eventually derived an explanation. When the islands arose as volcanoes emerging from the ocean, no animals or plants were on them. As the islands cooled and soils formed on the surface, a few animals migrated there from the mainland, and plant seeds made their way there as well—carried by animals or wind, or drifting on the water to the islands. With little or no competition from already established animals and plants, these organisms colonized the islands, and their descendants gradually spread to other parts. Variations among them allowed different types to adapt to different conditions. What once was a single species radiated into multiple species more rapidly on the islands than on the mainland because competition from already established animals and plants on the islands was minimal. In

The Origin of Species, Darwin explained this concept as "colonisation from the nearest and readiest source, together with the subsequent modification and better adaptation of the colonists to their new homes."[4]

On his return to England in 1836, after nearly five years on the *Beagle*, Darwin consulted with other naturalists to help him decipher the enormous amount of information he had accumulated on the voyage. He read vast amounts of information and conducted a wide variety of experiments. By then, he had rejected the notion that species were fixed and had instead adopted the view, already held by many, that new species had evolved from ancestral types.

Darwin's great discovery was not the idea of evolution—that was already well established—but instead one of its principal mechanisms. He postulated that new species arose through *natural selection*, which he succinctly described in *The Origin of Species* as this: "Any variation, however slight and from whatever cause proceeding, if it be in any degree profitable to an individual of any species . . . will tend to the preservation of that individual, and will generally be inherited by its offspring. The offspring, also, will thus have a better chance of surviving, for, of the many individuals of any species which are periodically born, but a small number can survive. I have called this principle, by which each slight variation, if useful, is preserved, by the term of Natural Selection."[5]

Darwin's theory of natural selection can be summarized as a series of logical points that build on previous points:

- In nature, each species produces more individuals than can survive in its environment.
- Because the environment cannot support all individuals, they compete with one another, and with other species, to survive and reproduce.
- They vary in their physical characteristics and those variations are, for the most part, inherited.
- Those whose physical characteristics offer them the best adaptation to their environment are the most likely to survive and reproduce.
- Those who survive and reproduce transmit these adaptive characteristics to their offspring through biological inheritance.

- The process continues generation after generation with gradual inherited change in a species resulting in increased adaptation to environments.
- Over many generations, the changes resulting from natural selection are sufficient to produce new species.

Darwin published *The Origin of Species*, a four-hundred-page book (which he called a "brief abstract"). The book was an enormous success, the first edition selling out on the day of its release. He immediately made some revisions and published a second edition. Over a period of years, he revised it through six editions, and it now stands as one of the most influential and bestselling books of all time.

It also was controversial from the day it was published, and remains so today, because it rejected the prevailing view that species are fixed—incapable of changing beyond certain boundaries. Many scientists accepted Darwin's claims while others remained skeptical. Today, however, the overall concepts of natural selection and the evolution of new species have been well documented through exhaustive observation and experimentation. Although some of the details Darwin proposed are now outdated, his overall theory stands on solid scientific ground.

The major piece missing from Darwin's theory when he presented it in *The Origin of Species*, and in his subsequent books, was an accurate understanding of inheritance. Unbeknownst to Darwin, at the time he published his book, a Moravian friar named Gregor Mendel was in the midst of experiments with peas that would reveal the process of heredity, which would be the key to explaining Darwin's theory. Darwin knew nothing about Mendel, but Mendel clearly was aware of Darwin. By the time Mendel presented his experiments in 1865, he had read *The Origin of Species* and was quite familiar with Darwin's theory.

Mendel made it clear that he viewed his own work as an important key for understanding evolution. He wrote on the first page of his paper, "It requires indeed some courage to undertake a labor of such far-reaching extent; this appears, however, to be the only right way by which we can finally reach the solution of a question the importance of which cannot be overestimated in connection with the history of the evolution of organic forms."[6]

Even though they were contemporaries, Darwin and Mendel never corresponded. Some have speculated that a "meeting of the minds" between the two might have furthered the world's understanding of evolution immensely. Others counter that Darwin's views were so different from Mendel's that the two might not have reconciled them.[7]

The publication of *The Origin of Species* catapulted Darwin to instant fame. Mendel's fate was the opposite. Throughout his life, no one recognized the importance of his work (not even Mendel himself), and his experiments, known only to a few people at the time, were mostly forgotten. Mendel died in 1884, highly respected as a priest but unknown as a scientist. Then, sixteen years later, three botanists independently conducted experiments on inheritance in plants and arrived at the same conclusions as had Mendel, only to find out that Mendel had discovered the keys to inheritance thirty-five years earlier. Mendel posthumously earned his fame in 1900 with the dawn of a new century.

Ironically, although Mendel perceived his work as a key to evolution, his newfound fame initially eclipsed Darwin's theory. For some, Mendelism seemed to contradict Darwinism—there was no way to reconcile the two. It was not until the 1930s that Darwin's theory of natural selection and Mendel's theory of inheritance were fully integrated during a period science historians have dubbed "the modern synthesis." Not surprisingly, Dobzhansky (the Russian biologist mentioned at the beginning of this chapter) was one of those who explained the connection. Darwinian principles of natural selection informed by Mendelian principles of inheritance are now known as *neo-Darwinism*, which is the foundation of the modern theory of evolution.

One key component was still missing from neo-Darwinism—the substance of heredity remained unknown. Most scientists were convinced it had to be protein. There were some indications that protein was involved in heredity—the word itself means "of first importance." But in 1944, experiments conducted by Oswald Avery and his collaborators, and later experiments published in 1952 by Alfred Hershey and Martha Chase, excluded protein and conclusively demonstrated that the substance of inheritance was DNA.[8] As methods for analyzing DNA developed over the ensuing decades, researchers could independently test the patterns of evolution derived from fossils and anatomy by examining the DNA of living organisms and, recently, the DNA from a few extinct species.

These studies repeatedly confirmed evolutionary histories developed through traditional means and opened a floodgate of new evidence revealing in stunning detail how organisms evolved. DNA studies of human evolution soon rose to the forefront. In the words of the scientists who sequenced and assembled the chimpanzee genome, "More than a century ago, Darwin and Huxley posited that humans share recent common ancestors with the African great apes. Modern molecular studies have spectacularly confirmed this prediction. . . ."[9]

The twenty-first century ushered in a quantum leap in our understanding of human evolution. Genomes are the complete DNA sequences of organisms, like enormous volumes of genetic instructions that determine how organisms develop and function. And each species has its own particular genome. The first draft of the human genome's DNA sequence and assembly was published in 2001,[10] with an updated, refined version in 2004.[11] The following year, the first draft of the chimpanzee genome was published,[12] then the rhesus macaque genome in 2007.[13] These three genome sequences offered the opportunity for scientists to compare the genetic information of three related species on an unprecedented scale. In essence, they could study the details of evolution across literally billions of bits of information in DNA and determine in the finest detail how our genome evolved. These extraordinary advances coincided with major fossil discoveries that significantly augmented the fossil record of human evolution. *It can now be justifiably said that we have more evidence of human evolution than for evolution of any other species.*

This book unites from different sources some of the most powerful information about how humans evolved and how we continue to evolve—from fossils, anatomy, archaeology, geography, and DNA. Nonetheless, for some people, the overwhelming evidence of human evolution is difficult to accept. Not on the basis of science, however, for the evidence consists of solid, irrefutable, and abundant observations. Instead, religious beliefs that seem to contradict evolution make the mere mention of human evolution uncomfortable, even offensive, for some.

In spite of massive amounts of evidence confirming our evolutionary history, many people refuse to accept the obvious conclusion that we evolved. A Gallup poll conducted most recently in December 2010 asked a group of respondents in the United States the following question: "Which of the following statements comes closest to your views on the origin and development of human beings? (1) Human beings have developed over millions of years from less advanced forms of life, but God guided this process. (2) Human beings have developed over millions of years from less advanced forms of life, but God had no part in this process. (3) God created human beings pretty much in their present form at one time within the last 10,000 years or so."[14] Forty percent of respondents selected option 3, an outright rejection of human evolution. This percentage, although disturbingly large given the scientific evidence of human evolution, represents a decline from the nine previous polls conducted from 1982 to 2008, which averaged 45 percent. In spite of this recent decline, a significant minority of Americans (four of every ten, on average) consistently rejects the very idea that we evolved.

Seizing on this rejection, those who embrace the creationism and intelligent design movements have promoted what they claim is scientific evidence disproving evolution. The publicity they have generated is immense. A host of books, television and radio shows, and Internet sites attack evolution while offering no scientifically valid alternative. The two most prominent creationist organizations are the Institute for Creation Research (http://www.icr.org) and the Discovery Institute (http://www.discovery.org), both of which offer detailed information about their causes on their websites.

Scientists have not been silent. Dozens of books, hundreds of articles, and well-organized online information, combined with high-profile court decisions, have thoroughly demonstrated that creationism and intelligent design *are not science*. One of the best scientific websites responding to creationist claims is the TalkOrigins Archive (http://www.talkorigins.org). The National Center for Science Education (http://ncseweb.org) serves as a source of information on attempts to require inclusion of creationism in public science education, and it has vigorously opposed such attempts. And the Public Broadcasting Service (PBS) offers an excellent resource on evolution at http://www.pbs.org/wgbh/evolution.

Driving the creationism and intelligent design movements is a perceived dichotomy between evolution and religion—to accept one is to deny the other. There is no question that the so-called young-earth creationist view—that the earth is less than ten thousand years old and everything on it was created pretty much as it is now—is absolutely incompatible with science.

A large number of scholars and scientists, however, have argued that when religious texts are viewed metaphorically rather than literally, science and religion can be viewed as complementary rather than hopelessly incompatible. I will not spend time here debunking creationism and intelligent design or discussing the compatibility or incompatibility of science and religion. Other books, including one of my own, have amply done so.[15] Instead, this book will focus on science, succinctly uniting some of the most extraordinary and overwhelming evidence of our evolution.

CHAPTER 2
EVIDENCE FROM OUR BODIES

S ome years ago, I lived in rural Virginia in a home surrounded by forest. Each spring, as the weather grew warm, insects emerged—literally millions of them, flying through the air; crawling on the ground; skirting about in the water of ponds, streams, and puddles. Butterflies visited blooming azaleas in our yard, hundreds of moths landed on our window screens at night, and fireflies put on magnificent light shows at dusk after a rain. Flying insects often took to the air just as the sun was setting. So did dozens of bats, silhouetted against the fading light of the evening sky as they darted about, eating the insects in mid-flight. I often watched, astonished, as these winged mammals performed extraordinary aerobatics—skillfully flying through the trees without ever striking a branch, diving downward to within inches of the ground before turning skyward in an instant, all in near darkness.

Bats are able to perform these spectacular feats because of their highly advanced sense of hearing. Each bat emits bursts of sound at pitches higher than our ears can detect. Imagine how a quiet evening would be transformed into one teeming with sound if we could hear them. Bat brains sort out the echoes that bounce off objects as minuscule as a mosquito or as large as a cliff side. They can follow the trajectories of flying insects and intercept them as morsels for their evening meals, all the while avoiding stationary obstacles or even moving objects, such as other bats in the vicinity. Remarkably, each bat can sense echoes from the sounds it emits and can sort them out from the cacophony of sounds emitted by other nearby bats, all within fractions of a second.

We humans marvel at a bat's ability to navigate with its ears largely because we are utterly incapable of doing so. Bats have certain specializations in their face, ears, throat, and brain that allow them to emit, detect, filter, and

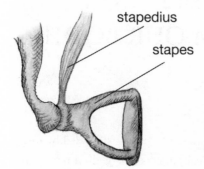

stapedius

stapes

**Figure 2.1. The stapedius muscle and
stapes bone in the mammalian ear.**

interpret sound in ways well beyond what we can do. For example, bats have a remarkable stapedius, a tiny muscle in the middle ear of mammals attached to the stapes, a minuscule stirrup-shaped bone (only about three millimeters long in humans), which transmits vibrations to the inner ear and auditory nerve.

The stapedius stabilizes the stapes, preventing it from vibrating too greatly. People for whom the stapedius fails to properly function perceive some normal sounds as if they were extremely loud. In bats, the stapedius carries out essentially the same function as in humans, but in a much more refined way. Bats contract and relax this muscle to help them distinguish nearby from distant echoes. The muscle, and the part of the brain that controls it, have become so refined in bats that the contraction-relaxation cycle can be incredibly rapid, as fast as two hundred times per second.

Evolution has treated bats well with an extraordinary set of adaptations integrated with one another to give them an acute ability to hear and interpret sounds in ways we cannot even imagine. Some of those adaptations are external, such as ears and faces contorted with complex protrusions of cartilage to capture and focus sound waves. Evolutionary biologist and artist Ernst Haeckel made thousands of exquisite drawings of organisms, many of them in his book *Kunstformen der Natur* (Artforms of Nature).[1] Haeckel's drawing of bat faces is one of the finest and expertly illustrates the wide range of ear and face adaptations.

The ability of bats to navigate by sound, which we find so astonishing, offsets their sometimes terrible eyesight. As Haeckel's drawing shows, many bats have small eyes, inherited from their evolutionary ancestors. The old adage

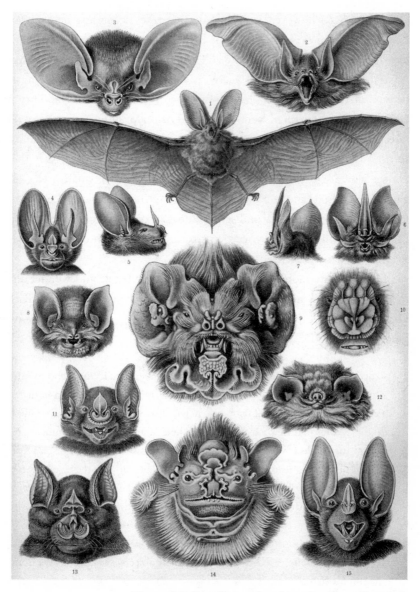

Figure 2.2. Drawing of bat faces by Ernst Haeckel.

"blind as a bat," however, is only somewhat true, depending on the species. Natural selection explains why some bats have lost visual acuity while others have retained it. Large fruit-eating and nectar-collecting bats of southern Asia, Australia, and the Indian and Pacific Islands—sometimes called flying foxes because of their large size—have no need to track flying insects. They have large eyes that serve them well, with acute vision for flying in daylight. Their faces look much like the face of a fox (hence the name *flying fox*), with no contortions to enhance sound capture. By contrast, those species of bats that do not see well are members of a large and diverse group of small insect-eating bats, derived from ancestors who spent most of their time in the dark, where vision is of little value. Natural selection has instead favored and refined their hearing, allowing them to succeed as creatures that exploit dark cave environments and chase insects in the night sky.

Suppose for a moment that these nearly blind bats could wonder among themselves about our extraordinary ability to navigate by sight. Richard Dawkins, in an entertaining passage from one of his books, mused on this same thought:

> I can imagine some other world in which a conference of learned, and totally blind, bat-like creatures is flabbergasted to be told of animals called humans that are actually capable of using the newly discovered inaudible rays called "light" . . . for finding their way about. These otherwise humble humans are almost totally deaf (well, they can hear after a fashion and even utter a few ponderously slow, deep drawling growls, but they only use these sounds for rudimentary purposes like communicating with each other; they don't seem capable of using them to detect even the most massive objects). They have, instead, highly specialized organs called "eyes" for exploiting "light" rays. The sun is the main source of light rays, and humans, remarkably, manage to exploit the complex echoes that bounce off objects when light rays from the sun hit them.[2]

Often we speak of *seeing* objects with our eyes. In fact, we are incapable of directly seeing any object. Instead, our eyes detect the light reflected from objects, our brains then interpret electrical impulses generated by cells in the retinas of our eyes, and these impulses are transmitted through nerves to our brains, where we visualize the objects. So well tuned is the human brain to interpreting visual images captured by our eyes that it can judge the size, distance, and even surface details of objects. Moreover, our eyes detect minute differences in the wavelengths of light, even mixtures of different wavelengths, which our brains interpret as variations in color.

Inextricably integrated with the ability of our eyes to detect objects by the light reflected from them is the ability of our brain to interpret visual signals in ways that go well beyond simple detection of light. For example, the image below consists of ink splotches on white paper (or, if you are reading this electronically, splotches of darkened black pixels against a background of white pixels on a light-emitting screen). Our eyes perceive light from the background as white and the lack of light from the splotches as black. However, there is no immediately recognizable pattern in the splotches, and they appear flat, an abstract arrangement with no apparent three-dimensional quality to them.

Now look at the same splotches rearranged into an assemblage that our brains instantly interpret as an image of Albert Einstein.

As straightforward and routine as this recognition is, something truly extraordinary is happening within our brains. Notice that Einstein's left cheek (his left, not yours) and much of his left ear are depicted as white space with no lines defining their boundaries. Yet our brain tells us exactly where the boundaries are, almost drawing the lines within our minds. The empty space does not appear abnormal or confusing.

Even more extraordinary in this exercise is our natural ability to envision an image that is completely flat as three-dimensional. We inherently know that the image is two-dimensional, and we have no trouble distinguishing it from a true three-dimensional depiction, such as a piece of sculpture. At the same time, thanks to our brains, we sense the image of Einstein as three-dimensional, unlike the purely two-dimensional image of fragmented splotches.

Our acute visual perception is a combination of eye structure and highly developed parts of the brain that interpret nerve signals sent from the eye, much as a bat's ability to echolocate (detect objects based on sound waves echoing from them) is coordination of its highly developed ears and brain. Another example of how our brains integrate visual signals is our ability to perceive depth. Our two eyes are located in the front of our heads, and they've been that way for quite some time—our primate ancestors had forward-facing binocular vision, as do our current primate relatives. Each eye, however, perceives a slightly different image. Focus your vision through both eyes on a distant object, such as something through a window or on the far side of a room. Then, place an upheld finger at arm's length in front of your eyes while still focusing on the distant object. You should see two images of your finger. While still retaining your line of sight, change your focus to the finger, and it becomes one image, while the object in the background becomes two. Try the same experiment with just one eye open and the objects remain single regardless of where your focus is. This double vision conferred by our two eyes is with us constantly, and our brains interpret it as depth perception, but we hardly notice it except in unusual situations, as when doing this sort of exercise or viewing a 3-D movie.

The brains of animals with eyes on the sides of their heads, such as rabbits, integrate their wide field of vision in ways difficult for us to imagine because our brains have evolved to interpret the images captured by forward-facing eyes. Even more difficult to imagine is how an insect brain interprets the multitude of images captured by its multifaceted eyes.

Although our brain makes it seem that we have sharp vision, in reality only a very small part of our vision is sharp. In the center of the retina (the back part of the eyeball's inside) is a small region of cells called the central fovea, which is where we have our sharpest vision. As you read this paragraph, close one eye and focus on a single word in the center of the page. While holding your focus on that word, try to read surrounding words without moving your focus from that single word. Those words close to your center of focus should be readable, but those even a few lines away are difficult or even impossible to identify without moving your center of focus to them. The region of your retina outside the central fovea is detecting the light from the entire page, but the lower concentration of light-detecting cells in that region, as well as a network of nerves and blood vessels on the surface of your retina, prevent you from readily identifying the words.

These nerves and blood vessels on the surface of your retina connect to light-sensitive cells. The nerve fibers form a network that coalesces in a bundle at the point where the optic nerve leaves the eye. There are no light-sensitive cells in the part of your retina where these nerves coalesce, leaving you with a fairly significant blind spot in each eye at this point. You don't perceive it as blind, however, because the field of vision for one eye overlaps the blind spot of the other, and, even with only one eye open, your brain compensates by filling in the blind space. At the end of this paragraph is a diagram that allows you to detect this blind spot in your right eye. With your left eye closed, hold this book at arm's length with your right eye focused on the X in the diagram, then pull the book toward you until the spot to the right of the X disappears. It should happen when the page is at a distance of about six to eight inches from your eye.

Notice as you do this blind-spot test that the color of the area where the spot vanishes appears white. Now look at the image below, which is identical to the previous one except the background is gray. Again focus your right eye on the X, and move the book until the spot disappears, paying attention to the color of the background where the spot vanishes.

This time the blind-spot region appears gray, like the rest of the background. Your brain fills the region in your field of vision where the blind spot is located with the surrounding color so that the blind spot does not distract you.

MUCH OF OUR ANATOMY MAKES SENSE ONLY "IN THE LIGHT OF EVOLUTION."

The point of these optical exercises is to exemplify how our brains have coevolved with our sensory organs and how our brains help compensate for their shortcomings. Remember Theodosius Dobzhansky's famous line "Nothing in biology makes sense except in the light of evolution." Nowhere is the validity of this statement more evident than in some strange features of our anatomy that make sense only from an evolutionary perspective. As quite a few biologists have pointed out, our highly developed eyes have some basic design flaws—the blind spot is but one of several. Another flaw, which is the reason for our blind spots, is the strange inside-out surface of our retinas. Most everywhere on the light-detecting surface of our retinas is a network of nerves and blood vessels.

The network of tiny blood vessels is unique to each person, allowing retinal scans to identify an individual even more readily than a fingerprint.

Why is this network on the surface of the retina, where it interferes with our vision, instead of behind it, where it could easily serve the retinal cells but be out of the way? According to the fossil record, as well as evidence in DNA, invertebrate animals in the oceans were the ancestors of the first vertebrates. Evidence from current marine invertebrates, such as squids and octopuses, tells us that our distant invertebrate ancestors probably had eyes with retinas that were right-side out, with the nerves located *behind* the light-detecting retinal cells, where they do not interfere with visual perception. Early in vertebrate evolution, before the first fish evolved, the retina apparently reversed itself to accommodate evolution of the rubbery lens in vertebrate eyes.

Several authors, among them Lewis Held Jr., Jerry Coyne, and Richard Dawkins, have explained how the vertebrate eye evolved, and I won't take the time here to repeat their detailed and excellent descriptions.[3] Suffice it to say that an optical instrument engineer could have easily redesigned the nerve network and its coalescence in the eye to make our retinas more efficient and to avoid making blind spots. Our evolution has instead compensated for the network of nerves and the blind spots by taking what was there and preserving those variants that functioned, allowing us to have excellent vision in spite of the flaws. As we saw a moment ago, our brains mask the blind spots so we hardly ever notice them, even though they are constantly there. Throughout the animal kingdom, brains have coevolved with the structures they control, allowing animals to adapt to their particular environments.

Countless other examples of anatomy explainable by evolution can be found throughout our bodies, some of them exacerbated in humans. One of the most obvious and problematic is the shape of the human pelvis, which makes human childbirth excruciatingly difficult and painful. Our distant reptilian ancestors— and most modern reptiles and birds—reproduced by expelling eggs through a central opening in the pelvic bones. Mammals have retained this pathway, giving birth through the same pelvic-bone opening that reptiles and birds use for egg laying. For nearly all mammals, including our closest primate relatives, the process is quite fast and efficient. But our large skulls, which evolved to house our large brains, have made it exceptionally difficult for a human infant's head to pass through the pelvis during birth. The infant must twist and turn as its head emerges, the immature skull misshaping itself to squeeze through the bony pelvis. The mother's pelvic bones, in turn, must partially separate by stretching a fibrous connection between them called the symphysis pubis.

The fossil record shows that in our recent evolutionary ancestors this pelvic opening gradually became larger to accommodate an increasingly larger skull. But its expansion has not kept pace with the expansion of our skulls. The human pelvic constriction is one of the most dangerous features of our anatomy. Throughout human history, complications of childbirth have been and still are a leading cause of death to both women and babies. As Lewis Held Jr. pointed out in his book *Quirks of Human Anatomy*, a human birth canal diverted to the front of the abdomen, such as through the navel, instead of through the pelvis, makes much more sense from a design standpoint. In a strange irony, our highly evolved brains have allowed us in recent times to divert birth through the abdomen. A cesarean section (commonly known as a C-section) is a surgical procedure that extracts the infant through the soft tissues of the abdomen to avoid forcing it through the narrow pelvic opening.

Our large skull is a feature exacerbated in humans, a consequence of the selective advantage our large brains give us. The excruciatingly painful and dangerous ordeal women and infants must endure during childbirth is a relic of our evolution.

WE SHARE MUCH OF OUR ANATOMY WITH OUR EVOLUTIONARY RELATIVES, BUT WITH ADAPTATIONS.

Let's return for a moment to bats. Most species of insectivorous microbats (the small bats that fly around at dusk eating airborne insects) have very poor eyesight, far less visual acuity than we do, yet the basic structure of their eyes is very similar to ours. By contrast, their ability to hear and interpret sound waves is vastly more refined than ours, even though the anatomy of their middle ears is likewise similar to ours. As is typical of mammalian ears, we and bats have three tiny bones in the middle ear—the stapes, the incus, and the malleus, sometimes respectively called the stirrup, anvil, and hammer because their shapes resemble those objects. These three bones connect at tiny joints and operate with one another as a unit, fully separated from the other bones in mammalian heads.

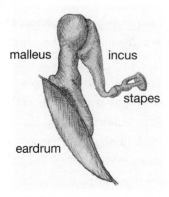

Figure 2.3. Bones in the mammalian middle ear.

Reptiles, on the other hand, have a single bone in the middle ear—the stapes. Where are the other two bones? As shown in figure 2.4, they are also present in reptiles but are called the quadrate (corresponding to the incus) and the articular (corresponding to the malleus). The quadrate is a bone at the base of the cranium (the upper part of the skull that houses the brain), and the articular is in the jaw, both bones joining each other where the jawbone meets the cranium. The quadrate connects the stapes to the cranium, and through the cranium to the rest of the skeleton.

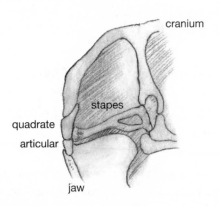

Figure 2.4. Ear bones in a reptilian skull viewed from behind in cross section.

This arrangement allows reptiles to readily hear low-pitch sound vibrations in the ground—such as the footsteps of an approaching predator—transmitted through their feet and the skeleton to the stapes, and through it to the inner ear. We, with our tiny middle ear bones free of the cranium and jawbone, sense low-vibration sounds in the ground not nearly as well as many reptiles, but we're much better equipped to hear higher-pitched sounds transmitted in the air.

How did this different positioning of bones in mammals and reptiles arise? Fortunately, the transition, as mammals evolved from reptiles, which took place from about three hundred to one hundred million years ago, is exceptionally well preserved in the fossil record, and paleontologists have documented how bone structure changed during that transition. In ancestral reptiles, before any mammals were present on the earth, the quadrate and articular bones initially composed the joint between the jaw and the cranium. Then, as the lineage leading to mammals evolved, these bones gradually became smaller. Eventually they were so small that they no longer functioned as a jaw joint but retained their original reptilian connections—articular connected to the quadrate, which is connected to the stapes. The joint between the tiny malleus (articular) and incus (quadrate) bones in our inner ear, which now transmits sound waves between those two bones, is the leftover joint of the reptilian jaw. The bones in their new positions, no longer needed for the jaw joint, have been co-opted for improved hearing, their exceptionally small size permitting mammals to hear high-pitch sounds well beyond the range that most reptiles are capable of hearing.

ANATOMICAL COMPARISONS AMONG LIVING ORGANISMS AND FOSSILS ALLOW US TO INFER HOW DIFFERENT STRUCTURES EVOLVED FROM COMMON ANCESTRY.

The comparison of bones in mammals and reptiles to determine how the mammalian ear evolved is an example of what scientists call *comparative anatomy*, and evidence from it tells us much about how a multitude of structures in our bodies evolved. The anatomical arrangements of bones, muscles, tendons, ligaments, nerves, organs, and many other features in different vertebrates are exceptionally similar, even when these structures carry out vastly different functions. For example, a horse has the same bones in its front leg as we do in our arm, wrist, hand, and middle finger. What we call the horse's front knee, for instance, corresponds to our wrist and has a collection of the same small bones our wrist has. The horse's hoof corresponds anatomically to the fingernail on our middle finger—their hooves and our fingernails are even composed of the same substances. The examples of such similar anatomical arrangements co-opted for different functions are countless. Other similar structures, such as our hearts, lungs, teeth, and eyes, carry out essentially the same functions in humans and in horses, albeit in somewhat modified ways.

We can best explain these similarities in anatomical arrangement with divergence in function if we infer that the similarities were inherited from distant common ancestry. Through many generations, inherited variations in one lineage were preserved through natural selection, resulting in accumulated changes so that they now serve a specialized function. Different variations in the same original structures were preserved in another lineage, and in this lineage they serve a different specialized function. Over many generations, the original structural *arrangement* is preserved in both lineages, but the structural *divergence* to serve different functions may be considerable.

When we compare our anatomy with that of other animals, both living and fossilized, we can see how anatomical features change through evolution. The process, as proposed by Darwin, is a consequence of natural selection. As he described it, "any variation, however slight and from whatever cause proceeding, if it be in any degree profitable to an individual of any species . . . will tend to the preservation of that individual, and will generally be inherited by its offspring."[4] In other words, slight variations in anatomy, if they offer an advantage and are inherited, will be preserved in offspring. Over many generations, what was initially a slight variation can gradually change into a major variation as lineages diverge.

Darwin proposed that this idea of gradual change in anatomical structure through natural selection could explain even the most complex and refined structures, such as a bat's ear or a human's eye, even though their complexity makes it difficult for us to envision how evolution could shape such a refined structure. Opponents of evolution sometimes gleefully quote from Darwin's *Origin of Species*: "To suppose that the eye with all its inimitable contrivances for adjusting the focus to different distances, for admitting different amounts of light, and for the correction of spherical and chromatic aberration, could have been formed by natural selection, seems, I freely confess, absurd in the highest degree."[5] Darwin's seeming confession, however, is purely rhetorical. Shortly after this sentence, he wrote, "Reason tells me, that if numerous gradations from a simple and imperfect eye to one complex and perfect can be shown to exist, each grade being useful to its possessor, as is certainly the case; if further, the eye ever varies and the variations be inherited, as is likewise certainly the case; and if such variations should be useful to any animal under changing conditions of life, then the difficulty of believing that a perfect and complex eye could be formed by natural selection, though insuperable by our imagination, should not be considered as subversive of the theory."[6]

Darwin based his conclusions on observations and information available to him in his day. The power of any theory rests on its ability to *predict* what additional observations should show us. And Darwin's theory of natural selection predicts that species with common ancestry should have similar anatomical features, but that those features should vary as explainable adaptations. Moreover, although the fossil record is inherently incomplete, we ought to see in it evidence of the type of step-by-step change Darwin proposed.

As an example, let's look at the forelimb of reptiles, amphibians, birds, and mammals. Most species that fall into these groups are *quadrupeds*—they have two forelimbs and two hindlimbs, although the exceptions (whales, dolphins, snakes, and legless lizards, to name a few) are evolutionarily important. If we compare the anatomical structures of the forelimb, we see a common pattern emerge. Let's start with our own forelimb—the arm, wrist, and hand (see figure 2.5, page 43). Your upper arm contains a single bone called the humerus. At the shoulder, it has a ball joint that fits into a socket in the scapula (the shoulder blade). This ball joint allows you to extend your arm in a wide range of angles at the shoulder. Your forearm has two parallel bones, called the radius and the ulna. The ulna attaches to the humerus at the elbow with a hinge joint—it moves only back and forth. In your wrist is a collection of eight small bones, called carpal bones, that allows the wrist to bend in many directions. The hand is composed of five sets of radiating bones. The five metacarpals are enclosed within the hand, and attached to them are the pha-langes, which form the fingers. The overall pattern, from shoulder to fingers, is one bone attached to the scapula by a ball joint with a hinge joint on the other end, followed by two parallel bones, then a collection of small bones, and five radiating digits.

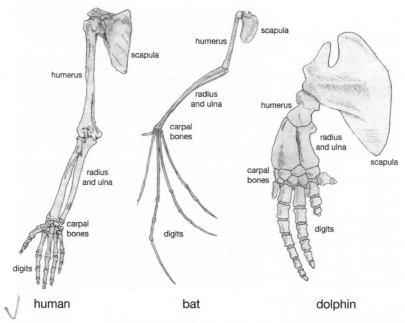

human   bat   dolphin

Figure 2.5. Comparison of human, bat, and dolphin forelimbs.

Now let's compare our forelimb with those of some other animals, starting with a bat. Even though a bat's forelimb serves a very different function than ours, it has the same pattern of bones. As shown in figure 2.5, a bat has a humerus with a ball joint on one end and a hinge joint on the other, radius and ulna, a set of carpal bones, and five digits, just as we do but in vastly different proportions. The same is true for a dolphin. It has the same pattern of bones that we have in its forelimb, all enclosed within a fin, where most of the joints are locked in position and barely function as movable joints. These five digits within a dolphin's fin are yet another example that makes sense only "in the light of evolution." We could go on, looking at lizards, frogs, dogs, cats, mice, and hundreds of other animals and find the same pattern.

But there are exceptions. Let's return to the horse's forelimb as an example. It has part of the pattern, but not all. A horse's forelimb has a humerus attached to the scapula by a ball joint, just as expected (see figure 2.6). However, the next bone, which, according to the basic pattern should be two parallel bones, is instead a single bone. Beyond that bone, at what people often call the horse's front knee, is a collection of small bones, called carpal bones, corresponding to those in our wrist. The rest of the leg contains the same bones as those in a single digit, and they correspond directly to the bones in the middle metacarpal and middle finger of the human hand. Finally, the hoof contains a bone corresponding to the final bone in our middle fingers. The outer part of the hoof is not bone but is composed mostly of a protein-based material, which is the same material in our fingernails.

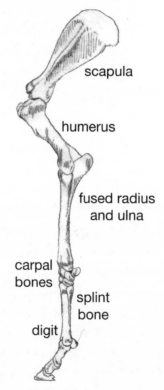

Figure 2.6. Bones in a horse's forelimb.

We can think, quite correctly, of horses and other hoofed animals as running on their fingernails and toenails.

If we look closely at the bones in a horse's forelimb, we can find hints of how it evolved from the basic pattern. The single bone that should be two, the radius and ulna, has grooves in it right where they should be if the radius and ulna anciently fused into a single bone (see figure 2.6, page 44)—and natural selection can explain the fusion. The elbow joint in mammals is a hinge joint that can only move back and forth, but we can still rotate our forearm and hand because the forearm consists of two bones. This ability to rotate our forearm is a clear advantage in our anatomy, given the ways we use our hands. A horse's forelimb, by contrast, is better served without rotation—restricted to forward and backward movement only. The fused radius and ulna give the horse added strength in its front legs as they move forward and backward, bearing much of the body weight, as the horse runs, trots, or walks.

A close examination of the bones in a horse's front leg also offers a clue about the fate of the extra digits. On either side of the single bone below the horse's "knee" are two slender pointed bones, called splint bones, which serve no essential function (figure 2.6). They are the remnants of what were once full-fledged digits. Here is where fossils offer us a clear picture of what happened, and they are among the best preserved of any set of transitional fossils. Figure 2.7 (page 46) highlights just a few examples of the many forelimb fossils of ancient horse species that lived during the past fifty-three million years.

Going back in time to the earliest fossils recognizable as horses is a group called *Hyracotherium*, which lived in the Eocene period about fifty-three million years ago. Species of *Hyracotherium* were about as large as a medium-sized dog, and they had four digits with small hooves on the front leg, the missing digit corresponding to our thumb.

By the early Oligocene, about thirty-eight million years ago, *Mesohippus* was a larger animal than its ancestors, and its legs had to bear more weight. The earliest species of *Mesohippus* also had four digits on their front legs, but the sizes had changed. The digit corresponding to our middle finger was substantially enlarged to bear the extra weight, and the one corresponding to our little finger was so reduced that it no longer touched the ground and had entirely lost its function as a weight-bearing digit.

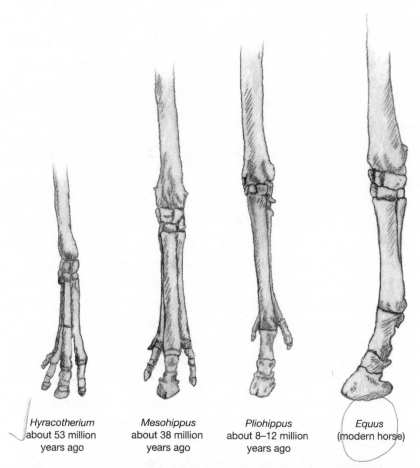

Hyracotherium
about 53 million
years ago

Mesohippus
about 38 million
years ago

Pliohippus
about 8–12 million
years ago

Equus
(modern horse)

Figure 2.7. Examples of forelimbs from ancient and modern horses.

From the middle Oligocene, about thirty million years ago, until the late Miocene, about five million years ago, the forelimbs of several different species of ancient horses had three digits—a large central digit and two smaller digits on either side, corresponding to our middle finger, index finger, and ring finger. As ancient horses became larger during this long period of geologic time, a larger central digit evolved to support the increased weight, while the flanking digits gradually became smaller. Initially, these flanking digits touched the ground and possibly stabilized the central hoof, but eventually they became so small that they lost contact with the ground and became ultimately nonfunctional relics, as in *Pliohippus*, which lived around eight to twelve million years ago.

By five million years ago, the anatomy of the modern horse had evolved. The two flanking digits were reduced to mere splint bones, hidden as vestiges below the skin, as in the horses of today. Donkeys and zebras also have these vestigial splint bones because their ancestral lineages diverged from the lineage that led to horses after the digits had become vestiges under the skin.

It is tempting to view the evolution of the modern horse as a straight line, yet a more accurate model is that of a branching tree. Ancestral types of horses diverged into branches, so that single-toed and three-toed horse species existed at the same time. Today, however, only the single-toed lineage has survived.

This discussion of forelimb evolution in the horse exemplifies two important aspects of evolution. First, a major change in structure typically requires a very long period of time. For example, it took more than forty million years for the single digit in the modern horse to fully evolve from four digits. Second, evolution often leaves *vestiges*, which are remnants of anatomical structures whose original functions are reduced or absent. The two splint bones in modern horses that flank the long metacarpal bone are vestiges of digits that in the evolutionary ancestry of horses were at one time functional for locomotion but no longer are.

VESTIGES AND ATAVISMS OFFER POWERFUL EVIDENCE OF HOW ANATOMY EVOLVES.

Examples of vestiges number in the thousands. A tiny sampling of the most obvious are remnants of sightless eyes—often buried under the skin—in cave fish and marsupial moles, small hindlimb bones in whales and some snakes, dew claws on dogs, and splint bones in horses, as seen in figure 2.8 (page 48).

These vestiges have no apparent function. However, an anatomical structure does not have to be functionless to be an evolutionary vestige. In some cases, a vestige may be a rudimentary structure that still carries out its original function to a substantially reduced degree, or it may have been co-opted for another function.

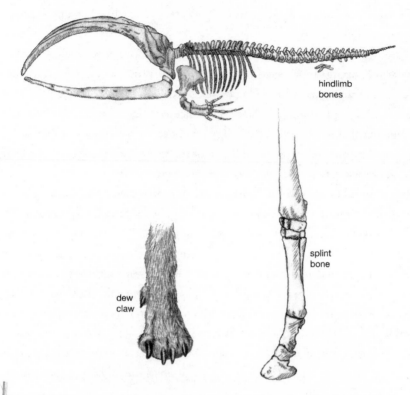

Figure 2.8. Three examples of evolutionary vestiges in animals: hindlimb bones in whales, dew claws in dogs, and splint bones in horses.

For example, ostrich wings are completely useless for flight but serve another function—ostriches flare them in mating rituals. The yolk sac is an embryonic structure that carries nutrient-rich yolk in the eggs of egg-laying vertebrates, such as most reptiles, amphibians, birds, and fish. The earliest mammals were egg layers, and the monotremes (platypus and echidna) are modern mammals that retain some of the characteristics of these primitive mammals, including egg laying. All other mammals, however, do not lay eggs, but their embryos still have a yolk sac, albeit with no yolk. Instead, in most mammalian species, the vestigial yolk sac has been co-opted as the site where the first blood cells are formed and transported to the embryo when the circulatory system first begins to develop.

Are there any evolutionary vestiges in humans? Our vestiges are not as strikingly apparent as sightless remnants of eyes, ostrich wings, dew claws, or splint bones, but we have many. We'll look here at just a few of the most obvious ones, as illustrated in figure 2.9. The yolk sac in human embryos mentioned a moment ago is an example. Goose bumps, which form on your skin when tiny arrector pili muscles in the skin contract, are vestiges of a time when our ancestors had hair covering their entire bodies. The function of goosebumps is to raise hairs to better insulate the animal when the air is cold, or to make the animal look larger and more menacing when frightened. We get goosebumps when we are cold or sometimes when we're nervous for these same reasons, although the hair they are supposed to raise is reduced to tiny hair shafts that do little to insulate the body or make it look menacing.

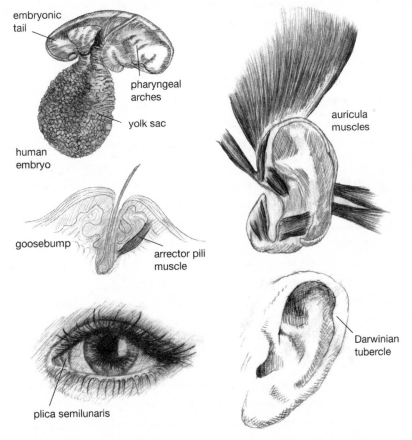

Figure 2.9. Examples of evolutionary vestiges in humans.

The inside corner of your eye has a small membrane called the plica semi-lunaris, which serves no function. In many other animals, this membrane is much larger and is called the nictitating membrane. It can move sideways to cover, moisten, and protect the eye, augmenting the function of the eyelid. For us, however, our eyelids must carry out these functions alone.

Human ears have auricula muscles attached to them; these are the same muscles that other animals use to raise their ears or move them backward and forward, as dogs, cats, horses, cattle, and deer do. Although a few people are able to move some of these muscles slightly, they are essentially functionless for human ears. Also, human and ape ears roll inward along the outer margin, which is not true for some monkeys. However, many people (and apes) have along the rolled-in portion of the ear a Darwinian tubercle—a small thickened point on the edge that is a vestige of a pointed ear (figure 2.9, page 49). The size of this tubercle varies considerably among people, ranging from very prominent to hardly noticeable.

Human embryos have numerous vestiges that gradually disappear as the embryo develops. The yolk sac we just discussed is one example. Another is the set of six pharyngeal arches that initially develop on each side just beyond the embryo's head (three are visible in figure 2.9). In fish, these same arches develop into gills and their associated structures, and our pharyngeal arches are the vestiges of gills in our distant fish ancestry. In humans, they have been co-opted as the precursors for other structures. All but the fifth arch (which gradually disappears) develop into anatomical structures of the head and neck, such as the incus, malleus, and stapes bones of the middle ear; the hyoid bone; and structures of the larynx (voice box), thyroid gland, thymus gland, and trachea; as well as several muscles of the head and neck.

Our embryos also have milk lines—lines of immature cells that are the precursors of mammary glands—running from the armpits to the groin. These lines are the sites in many mammalian species where two lines of mammary glands develop along the chest and abdomen for nursing litters, as in pigs, dogs, and cats. In humans, only two mammary glands typically develop, one on each milk line. In some people, the tiny hairs on their chests and abdomens form whorls along the milk lines on either side of the chest and abdomen as vestiges of multiple mammary glands in our distant mammalian ancestry.

About 5 percent of people have one or more additional nipples, called *supernumerary nipples*, along the milk lines, usually at the positions just above or below the typical nipples. Supernumerary nipples are often reduced in size, similar to the size of a mole, and some who have them mistake them for moles. In a few people, supernumerary nipples are larger and appear very similar to a fully developed nipple. In rare cases, as a young woman enters puberty, an extra breast may develop with a supernumerary nipple, a condition known as polymastia.

The word *vestige* is usually reserved for rudimentary structures that appear in nearly every individual of a species, whereas lost ancestral structures that occasionally reappear in a few individuals, such as supernumerary nipples, are called *atavisms*. One of the most striking examples of an atavism in humans is a true tail. Although many primates, such as monkeys, have tails, apes and humans lack them. During embryonic development, however, we all have true tails, with eight developing vertebrae in them. In nearly all human embryos, the final four vertebrae become reduced in size and disappear by the eighth

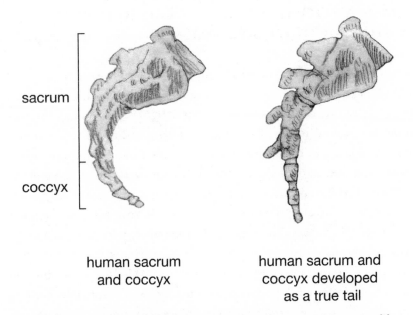

sacrum

coccyx

human sacrum
and coccyx

human sacrum and
coccyx developed
as a true tail

Figure 2.10. Comparison of the typical human sacrum and coccyx with those that develop as an atavistic true tail in humans.

week of embryonic development. The remaining four tail vertebrae do not fully develop and partially fuse during childhood and early adult development to form a small vestigial tail called the coccyx—which we often (appropriately) call the tailbone.

The coccyx has completely lost any tail-like function. It does not protrude from the body and now functions as an attachment site for several of our muscles. Occasionally, human infants are born with a small protruding tail that persists into adulthood if it is not surgically removed. Some of these are pseudo-tails, abnormal appendages of skin that can form nearly anywhere on the body but happen to form at the site of the coccyx. However, the majority are true tails, some containing the anatomical elements of a tail, including elongated vertebrae in the coccyx, cartilage, and, in some cases, muscles capable of moving the tail. These tails are typically short, with the same number of vertebrae as in a typical coccyx, but the vertebrae are more fully developed as tail vertebrae.[7]

HUMAN ANATOMY IS HIGHLY DERIVED FOR HABITUAL UPRIGHT BIPEDALISM.

There are many adaptations in our anatomy that allow us to carry out uniquely human behaviors. When humans and great apes are compared, some striking anatomical features clearly set us apart from the great apes. Among the most obvious are a considerable reduction in body hair; modified mouth and neck anatomy, allowing us to use spoken language; the length and positioning of thumb and fingers in our hands, permitting advanced dexterity and tool use; our enlarged brain and the cranium that houses it; and a host of adaptations that promote habitual upright posture and bipedalism—standing, walking, and running on two legs. The great apes—common chimpanzees, bonobos (a distinct chimpanzee species), gorillas, and two species of orangutans—have contrasting anatomical features, such as abundant body hair, the inability to speak, long fingers and a thumb position that are well adapted for grasping tree limbs, smaller brains and crania, and an overall anatomy that promotes a forward posture and four-limb knuckle or fist walking.

Because all great ape species have these and other similar anatomical characteristics in common, and because these features differ from those in humans, scientists for many years presumed that the great apes were more closely related to one another than any one was to humans. However, DNA studies consistently and repeatedly dispelled this presumption—the evidence overwhelmingly shows that the two chimpanzee species (common chimpanzee and bonobo) are more closely related to humans than either is to gorillas or orangutans.

Humans have diverged more from the human-ape common ancestral anatomy than have the great apes, manifested by the many unique anatomical features of our species. When examining anatomy, scientists often describe one form as *ancestral* and another as *derived*. A clear example is comparison of the human forelimb with the horse forelimb, which we discussed earlier in this chapter. The human forelimb (our arm) has the *ancestral* pattern with humerus, radius, ulna, wrist bones, and five digits. The horse forearm is *derived* from the ancestral pattern with humerus, fused radius and ulna, "knee" (wrist) bones, one functional digit, and two vestigial digits (the splint bones). Many of the anatomical features that are common to the great apes—abundant body hair, relatively small brain cases, and forward four-limbed posture, to name a few—were surely the anatomical features in the most recent common ancestor we shared with them and are thus ancestral. In the lineage leading to humans, many of these features were substantially modified—they are derived.

The fossil record of ancient humanlike species covering a period of seven million years has yielded considerable evidence about how and when these derived features evolved. We'll explore this evidence more fully in the next chapter. For now, let's focus on the derived adaptations that allow us to be habitually bipedal—to stand, walk, and run upright on two legs.[8] Although other primates may occasionally stand upright and walk bipedally, habitual upright bipedalism is a uniquely human trait. Even birds and extinct dinosaurs, which are and were habitually bipedal, have a much different form of bipedalism than we do, and it is a form that evolved separately and independently. Our form of bipedalism evolved exclusively in the human ancestral lineage and is not found in any other lineage—primate, mammalian, or otherwise.

The transition from the ancestral forward-leaning four-limbed posture to derived two-legged bipedalism required multiple derived anatomical adaptations. Let's review some of them from head to toe. The upper part of our skull, which is the entire skull except our jawbone, is called the cranium. At the base of the cranium, buried deep within the site where our neck meets our head, is an opening called the foramen magnum, where the brain stem exits the cranium and connects to the spinal cord. The human foramen magnum is positioned underneath the cranium, where it is better suited for upright posture. In great apes, it is toward the back of the cranium, consistent with the forward-leaning posture apes use to knuckle-walk.

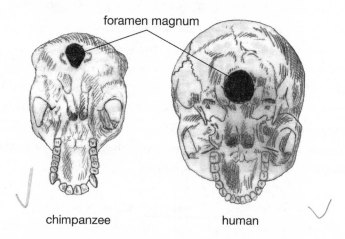

foramen magnum

chimpanzee   human

Figure 2.11. Comparison of how the foramen magnum is positioned in chimpanzee and human crania.

Figure 2.12 compares human and chimpanzee skeletons, with the human skeleton in an upright posture, the chimpanzee skeleton, first in the common knuckle-walking posture, and second in the posture chimpanzees occasionally assume when they stand and walk on two feet. The human vertebral column has a distinctive S-shaped curve when viewed in profile. The seven cervical (neck) vertebrae and twelve thoracic (rib cage) vertebrae curve toward the back, whereas the five enlarged lumbar (lower back) vertebrae reverse the curvature. As a consequence, our neck slants forward and our middle back curves outward behind us, whereas our lower back curves inward. The lumbar vertebrae are enlarged because they must bear the vertical weight of our upper body.

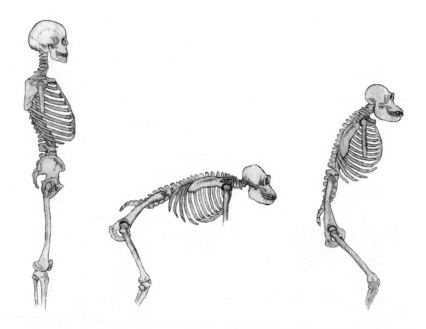

Figure 2.12. Comparison of human and chimpanzee postures. The upright human posture differs from the typical forward-leaning posture and more rare upright posture of chimpanzees.

By contrast, chimpanzees (and other great apes) have a much straighter vertebral column, with a slight overall backward curve, whether in knuckle-walking posture or standing upright. Chimpanzee lumbar vertebrae are proportionally smaller than ours because their body weight is distributed more evenly across the vertebral column when they assume their habitual four-limbed stance. When in an upright stance, chimpanzees bend their legs and lean forward to balance themselves, a stance that is consistent with the curvature of their vertebral column, but it is awkward for them. They typically do not maintain upright stances for long periods of time.

Our pelvis is one of the most highly derived parts for upright posture. It contains six bones, three on each side, two each of the ilium, ischium, and pubis (see figure 2.13, page 56). Our two ilia (plural of *ilium*) are the large curved bones we feel when we place our hands on our hips. They wrap around the back side of our hips in a half-bowl shape, which cradles the internal organs of our lower abdomen. The ilia are attached to the vertebral column at the sacrum, a triangular-shaped bone in the center of our lower back formed from five verte-brae that partially fuse during childhood and adolescence to form a single bone. The remnants of those original vertebrae are visible in the sacrum as protrusions pointing toward the back and as two rows of holes on either side of those protrusions. The coccyx, our vestigial tail, is attached to the lower part of the sacrum and usually contains four rudimentary and partially fused vertebrae.

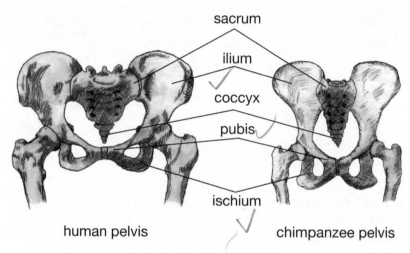

sacrum

ilium

coccyx

pubis

ischium

human pelvis   chimpanzee pelvis

Figure 2.13. Comparison of human and chimpanzee pelvises.

When compared proportionally, our ilia are shorter, broader, and more rounded than those of apes and monkeys, whose ilia are elongated and flat, shaped much like the blades of canoe paddles. As a consequence, our sacrum is positioned lower in the body, only slightly above the sockets where the femurs (thigh bones) attach to the pelvis. A chimpanzee's narrow, elongated ilia position the sacrum and coccyx higher above these sockets (see figure 2.12, page 55).

Not only our bones but also our muscles in the pelvic region are adapted for upright bipedalism. The most highly derived are the gluteal muscles, which give humans our distinctively rounded buttocks (figure 2.14). The enlargement and positioning of these muscles during the evolutionary transition to upright posture assisted our ancestors in maintaining themselves upright while walking and running.[9] They help us walk smoothly while maintaining our body in a straightforward position. You can feel the normal contractions of these muscles if you place your hands on the fleshy part of your buttocks as you walk. Notice that the contractions are most pronounced when one leg bears weight at the moment when you lift the other leg to step forward. Stop your walking while you have one leg raised, and you can feel the tensely contracted gluteal muscles on the other side keeping you from falling.

Our gluteal muscles are enlarged and repositioned compared to the same muscles in great apes (and other living primates). A chimpanzee, gorilla, or orangutan typically twists side-to-side when walking on two legs in part because its gluteal muscles lack the size, strength, and positioning to fully stabilize it while walking.

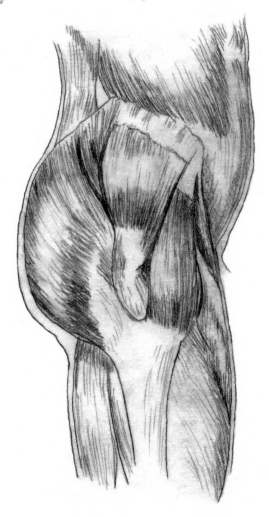

Figure 2.14.
Human gluteal muscles.

Our legs are proportionally longer than those of the great apes, giving us a greater striding distance for walking and running. The human foot, by contrast, is shorter and narrower, and it has highly derived adaptations for bipedalism. Our heel corresponds to a bone called the calcaneus, which in most quadrupedal animals is well above the ground and functions as part of a flexible, springy joint. A horse's hind leg, for example, has a backward-facing joint with a protruding calcaneus bone corresponding anatomically to our heel.

When we raise ourselves onto the balls of our feet to stand, walk, or run tiptoed, we mimic the way most animals stand, walk, and run, except for hoofed animals, which stand, walk, and run on the anatomical equivalents of our fingernails and toenails. But even they use their "heel" as a springy joint for walking and running. We, on the other hand, use our heels to impact the ground and rock our feet forward.

We share this feature with our great ape cousins, who not only walk

Figure 2.15. Bones in a human foot and horse's hindlimb.

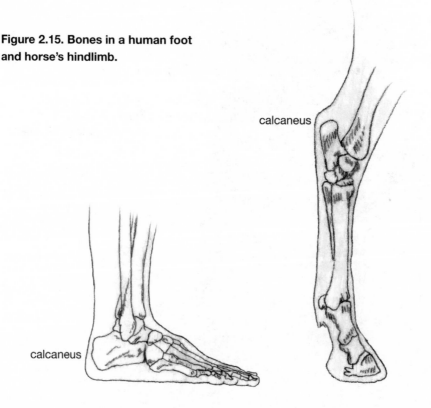

calcaneus

calcaneus

on their heels but press their heels against tree trunks and branches when climbing or resting in trees. As it turns out, our flattened foot and heel, which we probably inherited from a tree-dwelling ancestor, is efficient for walking but not necessarily for running. Plantigrade (heel-first) walking is more energy efficient than digitigrade (tiptoe) walking in humans, but the same cannot be said for running. As one group of researchers put it, "Relative to other mammals, humans are economical walkers, but not economical runners. Given the great distances hunter-gatherers travel, it is not surprising that humans retained a foot posture, inherited from our more arboreal great ape ancestors, that facilitates economical walking."[10]

The parallel, forward-pointing conformation of our toes also enhances our bipedalism. Among all living primates, we are the only ones whose large toe is parallel with the other toes, and our toes are relatively short. We are comically clumsy at grasping objects with our toes, especially when compared to apes and monkeys, whose long toes and thumb-like opposing large toe allow them to readily grasp objects with their feet, especially tree limbs when climbing. Our forward-facing and shortened toes are much better adapted to bipedal walking and running than are the feet of any other primate.

HABITUAL UPRIGHT BIPEDALISM HAS GIVEN US BOTH ADVANTAGES AND DISADVANTAGES.

As we will see in the next chapter, fossil evidence shows that bipedalism evolved very early in ancient humanlike species, well before most of the other uniquely human traits, such as tool use and the anatomical modifications necessary for speech. We might not intuitively see any connection between bipedalism and the evolution of tool use and speech; however, there is convincing evidence that they *are* related. Bipedalism freed the arms and hands from constraints imposed by quadrupedal locomotion. Adaptations that favor advanced tool use could best evolve only after bipedalism was habitual and had fully freed the arms and hands for other functions. This prediction is consistent with the fossil record—the first fossil evidence of tool use appears after the anatomical features favoring habitual upright bipedalism were well developed.[11]

Our arms are proportionally shorter than those of the great apes, who must be able to comfortably reach the ground in their quadrupedal stance as the vertebral column slants upward toward the head. Ape fingers are long and curved, adaptations well suited to tree climbing, swinging from branches, and knuckle-walking. Our fingers are shorter, and our thumb is positioned in such a way that we can easily touch the tips of each finger with it, giving us highly advanced dexterity. Our brains have acquired the ability to expertly control a practically infinite number of hand movements that today include such advanced and uniquely human feats as shaping pots, carving wood, weaving textiles, sewing, writing, typing, drawing, playing musical instruments, typing on a computer keyboard, and countless other intricate activities.

At first glance, it's difficult to see how the development of speech could in any way be tied to the evolution of bipedalism. However, a substantial body of research has shown that in vertebrates, especially in mammals, breathing is tied to locomotion, humans being the major exception.[12] When running, quadrupedal animals must coordinate their breathing with the motions of their gait because muscle contractions in their horizontal rib cages, which force breathing, are an integral part of the running motion. Horses exhale and inhale in rhythm with their gallop, which riders can readily hear when on a galloping horse. The same is true for apes as they move on all fours. Humans, however, have acquired the ability to uncouple breathing from running.

This ability offers several advantages, one of which is our freedom to consciously alter running speed without having to alter our breathing to match our steps. This is advantageous because people can coordinate breathing to meet oxygen needs rather than being confined to the rhythm of the gait. Speech requires highly controlled breathing, and a first essential step toward speech is the ability to control breathing without being forced to take breaths in step with movement. Once breathing was no longer constrained by locomotion, the other anatomical features necessary for speech were free to evolve.

In spite of the many advantages our habitual upright bipedalism offers us, it comes at a heavy cost. The vertical weight of our bodies takes a powerful toll on us. Our femur (thigh bone) and tibia (the larger bone in our calf) are enlarged at the knee, which also has thick cartilage padding between the two bones and strong ligaments holding them together at the knee to bear the vertical weight of our bodies and the strain of keeping them upright. Although a few million years of evolution have given us thickened and strong knees, one of the most common of human ailments is knee pain, and sometimes knee failure requiring surgical repair—a costly consequence of our bipedalism.

Although our lumbar vertebrae have become enlarged over evolutionary time in response to the extra weight they must bear because of our upright posture, this enlargement has not reached optimal efficiency, as anyone with lower back pain—also one of the most frequent human ailments—can confirm. This problem is especially acute for women in the later stages of pregnancy, whose lumbar vertebrae undergo exceptional strain when the added weight of the fetus, placenta, and fetal fluid shifts the center of gravity forward. The weight of a fetus in a pregnant great ape is more evenly dispersed and balanced across its vertebral column as it walks on all fours.

The shifted center of balance in pregnant women is not the only adverse effect of bipedalism on pregnancy and childbirth. For our ancestors to become habitually bipedal, the pelvis had to undergo some radical changes. The elongated pelvis of most species, including our great ape cousins, has an oval-shaped opening in the bones through which offspring pass during birth. Great ape females can give birth by themselves quite easily and quickly—and their young are more fully developed than human babies at birth. Our bowl-shaped pelvis, so essential for upright posture, has constricted this opening into an almost circular passage that is proportionally smaller in humans when compared to our great ape relatives. This smaller pelvic passage is at odds with the proportionally larger heads of human babies—an adaptation to accommodate our larger brains. As we noted earlier in this chapter, a human mother's pelvic bones must stretch apart where they connect at the pubis to allow the baby's head to pass during childbirth, one of several factors that prolong human childbirth and render it excruciatingly painful and dangerous to both mother and infant. As anthropologist Craig Stanford, who has made a career of comparing ape and human anatomy, put it, "Everywhere in the world human birth is an ordeal. Among our great ape relatives, however, birth is definitely not such an ordeal . . . the moment of birth is high speed compared with the human experience."[13]

The earliest evidence of emerging bipedalism may date back as far as five million years, which is but a blink of an eye in evolutionary time. Given that it took about forty million years for the horse's leg to evolve from four toes to one, it is quite remarkable that the numerous derived adaptations in humans have evolved so quickly. The human body is in many ways an efficient and fluid machine, so obvious in a skilled athlete, gymnast, figure skater, or dancer. It gives us many advantages, and our uniquely human adaptations have allowed us to become one the most dominant and adaptable species in the earth's long history.

But our bodies are far from perfect. Beyond some of the disadvantages our bipedalism confers on us, there are many other evolutionary flaws in our anatomy, such as excessive tooth crowding in a jaw that has grown proportionally smaller over recent evolutionary time, blind spots in our eyes, and an inherently higher probability of cancer than in our great ape cousins.[14] Moreover, evolution has left us with some traits that were beneficial in our ancient history but now promote serious health problems in modern times, such as cravings for a diet high in sugar, fat, and salt.

When did the traits that make us human evolve? Fortunately, the fossil record of our ancient humanlike relatives has in recent years become exceptionally rich, albeit still inadequate to answer certain critical questions. In these fossils, we see powerful clues about the evolution of bipedalism, brain expansion, emergence of tool use, and countless other uniquely human characteristics. The next chapter explores the ancient history of the *hominins*, our closest evolutionary relatives since the divergence of the lineages leading to humans and chimpanzees, and what their fossils tell us about our own evolution. Of these species—more than twenty-five now identified and probably others yet to be discovered—all suffered extinction, except for us.

CHAPTER 3
EVIDENCE FROM THE EARTH

A few years ago, I was hiking in a remote desert as the sun was setting, the sky filled with pastel hues of orange, pink, lavender, green, and blue. As I stood near a cliff in the fading light, mesmerized by the magnificent scene unfolding before me, I began rummaging through some broken pieces of shale at my feet. After a few minutes, I came across a trilobite fossil. Some more rummaging and I found another, then a third.

I had stumbled across a formation of sedimentary rock called the Wheeler Shale. It dates to a geological period known as the Middle Cambrian and consists of sediments deposited on an ancient shallow seafloor more than 500 million years ago, when most of what is now the United States was near the equator, a good part of it undersea. At that time, trilobites were among the most diverse and abundant organisms on the earth, much like insects are today. The first trilobites emerged around 550 million years ago. For the next 300 million years, a diverse array of trilobites evolved as others went extinct. The last trilobites to live in the earth's oceans died out during the great Permian extinction, about 250 million years ago, when only a fraction of all species on the earth managed to survive.

As I held those trilobite fossils in my hands, surrounded by the stillness of the desert and colored sky, a magnificent feeling of awe overcame me. I imagined these creatures scurrying along the seafloor, scavenging for food, avoiding predators, and seeking mates. At death, sediments covered their

Figure 3.1. A trilobite fossil.

remains, preserving them from complete decay. There they rested—silent, buried, and eventually petrified—for more than five hundred million years before another organism encountered them. I felt honored to be that organism.

HOW DO FOSSILS FORM?

Few people realize how abundant fossils are. Although I hunt for molecular fossils in DNA of modern organisms on a professional level, I enjoy hunting for true rock-solid fossils as a hobby. I've encountered them in North America, South America, and Europe. I've even found them in some seemingly unlikely places. When I visited the capitol in Richmond, Virginia, I noticed spiral-shell fossils in the polished stone-floor tiles of the rotunda, under the constant gaze of sculptor Jean-Antoine Houdon's magnificent statue of George Washington. Even the statue itself is a collection of fossil remnants. Houdon carved it in marble, a metamorphic stone formed when limestone buried deep within the earth was subjected to tremendous pressure and heat, gradually turning the opaque limestone into translucent, crystalline marble. The ancient limestone from which the marble was formed was originally a conglomeration of seashells and coral, crushed and compacted into stone by geologic forces.

Fossils are relics of the past, remnants of organisms that lived and died through eons of time. They offer us a small but incredibly informative sample of what life was like throughout the earth's history, as well as clues to how it evolved. Although fossils may be abundant, it is remarkable that we can find fossils at all, especially those of soft-bodied organisms. Only the tiniest fraction of organisms that ever lived were fossilized. When an organism dies, microorganisms immediately begin digesting its remains. The billions, even trillions, of microorganisms thriving on the remains of a dead animal or plant nearly always digest the material entirely, leaving no recognizable trace behind. Unless organisms have calcified shells, as do clams and oysters, or other rock-like body parts, they must die in a place where they are rapidly buried by sediments in order to be fossilized. This protected environment must delay decay for the fossil to form. Over immense periods of time, impressions of these organisms remain in the sediments as minerals fill the impressions and petrify.

For a fossil to become scientifically informative, however, it must make its way into the hands of a trained paleontologist who knows how to analyze and interpret information gleaned from the structures in the fossil, the sediments in which it was found, and its geographic location. We can safely conclude that most of the vast number of fossils on the earth are still buried, many of them inaccessible. For a fossil to be found, geologic upheavals and erosion must bring it to the surface where it is exposed, or where it is at least near enough to the surface to be accessed through digging. And, if a fossil is exposed, someone must find it and protect it before the forces of nature erode it away.

FOSSILS TELL A TREMENDOUSLY RICH STORY OF LIFE ON THE EARTH.

Countless fossils excavated, collected, and studied for centuries offer an overall picture of the earth's geologic history and of the evolution of plants, animals, and microorganisms throughout eons of time. The oldest undisputed fossils date to about 3.5 billion years ago and consist entirely of microscopic remnants of ancient bacteria. For more than two billion years, bacteria were the only known forms of life on the earth.

Fast-forward to 542 million years ago, and we arrive at the beginning of the Cambrian period, when large numbers of sea-dwelling animal species appeared, including the trilobites I found on my evening hike. Some people call this appearance of biological diversity the *Cambrian explosion* because a wide range of ocean-dwelling species appear in the fossil record. A more accurate term is the *Cambrian radiation* because, at that time, large numbers of species evolved over tens of millions of years, radiating from a much smaller number of ancestral species.

After the Cambrian, five major mass extinctions punctuate the history of life. The most serious of these was the Permian extinction, which happened 251.4 million years ago. Nearly all of life died during this extinction, leaving a relatively small number of surviving species to repopulate the earth. The cause of the Permian extinction has not been fully determined. Another major extinction happened 65.5 million years ago and is well-known because it coincides with the extinction of the last large dinosaurs. Its most probable cause is well documented—a large asteroid struck the earth near what is now the northern edge of the Yucatán Peninsula in Mexico. The massive climate change that resulted from debris jettisoned into the atmosphere when this asteroid struck probably caused this extinction, which eliminated 70 percent of the species alive at the time. This extinction is also important because it ended reptilian domination of the land and allowed a diverse array of mammalian species to evolve, including our earliest primate ancestors.

Certain fossils are so abundant and have been so thoroughly studied that newly discovered examples of them offer essentially no new scientific information. For example, the trilobite fossils I found were of the species *Elrathia kingi*, which is a common type of trilobite fossil that has been well studied. So many specimens of *Elrathia kingi* have been unearthed that people all over the world own them. They sell for a few dollars in gift shops and on the Internet, and often are made into jewelry.

Paleontology is now so advanced that scientists can use evolutionary theory to predict where they will find a particular type of previously undiscovered fossil. For example, according to the fossil record, the first tetrapods (vertebrate animals with four limbs) arose from lobed-fin fishes. However, before 2004, no one had discovered a good example of a fossil that was transitional between lobe-finned fishes and early tetrapods. The oldest lobe-finned fish fossils are found in sedimentary rocks dated at about 390 to 380 million years ago, whereas the oldest tetrapod fossils are dated to 363 million years ago. Examination of rocks bearing these fossils revealed that both types lived in shallow freshwater.

A team led by American paleontologists Neil Shubin of the University of Chicago and Ted Daeschler of the Academy of Natural Sciences in Philadelphia researched geological records for locations of exposed sedimentary rocks deposited in freshwater dating to between 380 and 363 million years ago as potential sites for finding transitional species between lobe-finned fishes and early tetrapods. They identified the Fram formation on Ellesmere Island in Canada as a site with the right criteria. It lies in a remote arctic region accessible for only a short period of time each summer. The team searched there for three summers and found interesting fossils, but not the transitional form they were seeking.

Then, during the fourth summer, they struck pay dirt, so to speak. The team found twenty specimens of a species with some features of lobe-finned fishes and others of early tetrapods—a true transitional form. The rock layers where they found this species were 375 million years old, the very time period when such a transitional form should have lived. They named the species *Tiktaalik*, which means "large freshwater fish" in the language of the local natives. Neil Shubin recounts the excitement of the find in his bestselling book *Your Inner Fish*.[1]

THE FOSSIL RECORD HAS OFFERED MUCH EVIDENCE ABOUT HOW WE EVOLVED.

What do fossils tell us about our most recent evolutionary ancestors—the human branch of life's family tree? Paleontologists have used essentially the same strategy as the group that found *Tiktaalik* to improve the likelihood of finding fossils of our closest extinct relatives, which, along with ourselves and the great apes, we call hominids. When Darwin wrote *The Descent of Man* in 1871, he predicted that extinct species of hominids had existed and that their fossils should be found in Africa. At the time, however, people searching for hominid fossils confined their efforts mostly to Europe, where only a few hominid fossils had been discovered. Bones of three Neanderthals (*Homo neanderthalensis*), a recent but now extinct relative of modern humans, had been found in Belgium, Gibraltar, and the Neander Valley of Germany, and bones of an extinct ape species, *Dryopithecus fontani*, had been discovered in France. The following year, fossils of another extinct ape, *Oreopithecus bambolii* (sometimes nicknamed the Cookie Monster because of its Latin name) were unearthed in Italy. This discovery later sparked fascination when scientists found that *Oreopithecus* was a bipedal ape, albeit with a different and independent form of bipedalism than that of humans.[2]

Sensationalism about human origins ran rampant for the next half a century, with a mixture of legitimate science and fraud. The most notorious fraud was the "discovery," beginning in 1915, of two specimens of a species dubbed Piltdown Man in an English quarry. These fossils were purportedly a transitional form between apes and modern humans that lived in ancient England. Over a period of three decades, as legitimate hominid fossils were discovered, paleoanthropologists had difficulties fitting the Piltdown fossils into a plausible evolutionary scenario with these other fossils.

In 1953, the reason for these difficulties became obvious—Piltdown Man was a fraud. Chemical dating methods showed that the skull specimens were from 620-year-old human skulls and that the single jawbone specimen was a 500-year-old bone from an orangutan. One tooth was from an elephant, another from a hippopotamus, and a third from a chimpanzee. Someone had treated the specimens with iron and manganese solutions to make them appear ancient, and the teeth had been filed to appear like fossil hominid teeth.

Unfortunately, the Piltdown fraud distracted people's attention from a remarkable and fully legitimate find. In 1924, a colleague delivered a box containing fossils to anatomist and anthropologist Raymond Dart, who recognized one of them as a hominid skull. This skull has been dubbed the "Taung child" because it was found near Taung in South Africa and still had its milk teeth.

This fossil was discovered in Africa, consistent with Darwin's deduction that Africa was the site for the origin of humans. All fossils of humanlike species discovered outside of Africa were quite similar to modern humans and, as dating methods would later show, were relatively recent, most dating to no more than a few hundred thousand years ago. The Taung child fossil turned out to be more than *three million* years old, and its discovery began a long series of discoveries in Africa of very ancient hominid fossils, dating back as far as seven million years ago. The fossil evidence that eventually accumulated from Africa decisively and unambiguously pointed to that continent as the place of origin for modern humans.

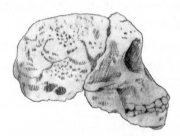

Figure 3.2. The Taung child. The front portion is a direct fossil of the skull, whereas the rear portion is an endocast of the brain case. The two portions fit together like pieces of a puzzle.

Not surprisingly, our three closest living evolutionary relatives—common chimpanzee, bonobo (an endangered species of chimpanzee with a limited geographic range), and gorilla—are also native to Africa—further evidence of an African origin for the earliest members of the human–great ape ancestral lineage.

It would be especially informative for us to compare fossils of extinct humanlike species with fossils in the great ape ancestral lineages. Unfortunately, the fossil record of most great ape ancestral lineages is exceptionally poor, and the reason for that is clear. Chimpanzees, bonobos, and gorillas currently live in the rain forests of Africa, as did their evolutionary ancestors. Rain forests are some of the *worst* places on the earth for fossil formation and discovery. In the hot, wet conditions of a rain forest, the dead bodies of mammals decay too quickly to be preserved as fossils, and even those rare fragments (such as teeth) that might have been preserved are practically impossible to find.

Even so, a few great ape fossils have been found, most of them recently. *Chororapithecus abyssinicus* is a species of great ape dating from 10 to 10.5 million years ago, fossils of which were found in Africa in the Afar rift of Ethiopia, and is thought to be related to the ancient ancestors of the gorilla.[3] Also, some fossilized chimpanzee teeth dating to 545,000 years ago were found in the Tugen Hills of Kenya.[4] Some ape fossils have been discovered outside of Africa. Earlier, we mentioned *Dryopithecus* and *Oreopithecus*, discovered in the nineteenth century in Europe. Also outside of Africa are ape fossils belonging to a lineage that led to the two species of modern orangutans, which are native to Indonesia and Malaysia. The ancestors of the orangutan lineage apparently migrated out of Africa through Asia about twelve million years ago, diversifying into a number of species. One of them, *Gigantopithecus*, identified mostly from fossil teeth and a piece of jawbone discovered in China, is the largest hominid known—it was about ten feet tall when standing upright (although it was probably a quadrupedal knuckle walker). The two modern Indonesian and Malaysian species of orangutan are the only remaining descendants of this lineage.

Accumulated information has allowed paleoanthropologists to systematically focus their attention on the most likely areas and geological formations in Africa for discovery of hominid fossils. The result has been an impressive number of discoveries unearthed at sites ranging from northern Chad to South

Figure 3.3. The Great African Rift Valley, where hominin fossils are most abundant.

Africa, most in the Great African Rift Valley, with an especially large number of specimens found near Lake Turkana in northeastern Kenya and the Afar region of Ethiopia.

The abundance of fossils in this area is a consequence of several factors. First, the word *rift* implies some sort of geological separation. The valley is in fact a rift system, where three tectonic plates—major masses of the earth's crust—have been moving apart and shifting relative to one another for millions of years. This tectonic activity has uplifted sedimentary layers, bringing them to the surface and exposing fossils.

Tectonic movements also produce volcanic activity, and radiometric dating methods allow scientists to readily date ash from volcanic eruptions.

As these eruptions happened repeatedly over millions of years, they deposited layers of ash that establish date ranges for fossils derived from the dates of the closest ash layers below and above the geologic layers where the fossils are found.[5]

Humanlike fossils dating back to the time when the human and chimpanzee ancestral lineages separated, between seven million and five million years ago, are now available. Sparse and fragmentary fossils are all we currently have of these earliest species. However, from about four and a half million years ago to the present, the fossil record is exceptionally rich with fossils from multiple individuals, some of which have partial to nearly complete skeletons.

For decades, scientists and nonscientists alike have used the term *hominid* in reference to humans and their extinct relatives. The great apes—chimpanzees, gorillas, and orangutans—by contrast were at one time called pongids. These terms, *hominid* and *pongid*, were derived from the scientific classifications Hominidae, the family to which humans belong, and Pongidae, once considered to be the great ape family. However, as information from DNA-based studies accumulated in recent years, it became increasingly clear that chimpanzees and bonobos are more closely related to humans than they are to gorillas or orangutans, nullifying any reason for separating the Hominidae and Pongidae. Humans and great apes are now classified into a single family called Hominidae, and thus humans and great apes, as well as their extinct relatives dating back to the common ancestor of humans and great apes, are all hominids.

For the remainder of this chapter, we'll use the term *hominin* to refer to species in the *hominin clade*, which includes humans and their extinct relatives that lived since our ancestral lineage diverged from the *panin clade*, the lineage leading to the common chimpanzee and bonobo (the two living chimpanzee species) and their extinct relatives. We'll use the term *hominid* to refer to members of the current family Hominidae, including humans and great apes and all our and their extinct relatives back to a common ancestor for all hominids. You should be aware, however, that some people still use *hominid* in its former sense, referring only to humanlike species as hominids, excluding the great apes, which can result in some confusion.

Momentarily we'll see what the hominin fossil record tells us about how our distinctly human anatomy evolved. Before we do so, however, we need to dispel two common misconceptions about human evolution—the so-called ape-to-human transition and the notion of missing links.

It is not uncommon to hear opponents of evolution utter something along the lines of "If chimpanzees are the ancestors of humans, why have we changed so much and they haven't?" We need to remember that modern apes clearly are *not* our ancestors. Instead, ample evidence—from fossils, from our shared anatomical and biochemical features, and from our DNA—shows that we and great apes descended from common ancestry. We should think of them as distant cousins, not as distant grandparents. We do not know which ancient species was the common ancestor of humans and chimpanzees—although one fossilized species, *Sahelanthropus tchadensis*, has been suggested as a possible candidate, or more likely, a near relative of it. Whatever that common ancestor was, it certainly had features common to both humans and great apes, and, as fossils of early hominins show, it had features now modified or no longer present in any of its modern descendants. Likewise, its modern descendants now have features it likely did not have. That it had many features of modern apes is certain because humans are the most anatomically derived species of all modern hominids. That common ancestor, if we could see it now, probably appeared very apelike, but it was not a chimpanzee.

The second misconception about human evolution (and evolution in general) is the notion of "missing links." Opponents of evolution often claim that for evolution to be true, the fossil record must show an ancestral "chain" from one species to another. They point to the obvious gaps in such a chain as missing links and claim that the missing links probably never existed in the first place; therefore evolution must be false.

When considering this notion, we need to first recognize what we mean by the term *species* and how species names are applied to fossils. As with living species, scientists assign Latinized scientific names to extinct species identified from fossils. The first name is the *genus* (plural *genera*), which designates a group of species that are highly similar and closely related, whether living or extinct. For example, the anatomical features of humans and Neanderthals are sufficiently similar for scientists to group them in the genus *Homo*. The second name is the species designation. Humans and Neanderthals have enough unique distinguishing features in their anatomy and in their DNA to be classified as separate species—*Homo sapiens* and *Homo neanderthalensis*. The same is true for the two living chimpanzee species. The Latinized scientific name for the common chimpanzee is *Pan troglodytes*, and for the bonobo, it is *Pan paniscus*.

Although scientists routinely assign Latinized scientific names to species, both living and extinct, we must remember that our current naming system is a convention invented by people. Our highly evolved minds seem to have an insatiable urge to categorize everything, and, over a period of centuries, scientists have devised an excruciatingly strict set of rules to categorize nature. But nature is under no obligation to follow our rules. We often describe a younger species as having evolved from an older one, as if the older species suddenly changed into the younger one in the blink of an eye. Evolution, however, is an ongoing process that sometimes proceeds slowly; other times, less slowly, but it is always under way. It is best to think of extinct species in the fossil record as snapshots taken at various points along an immense journey of slow but inevitable change. Species alive today are also snapshots of a point in evolutionary time. They, too, will evolve—or go extinct—as time passes.

Unless we had a complete fossil record representing every infinitesimally minor change, so-called missing links—or, better, gaps in the fossil record—are inevitable. As Richard Dawkins once quipped in a televised interview with an opponent of evolution, "Every time a fossil is found which is in between one species and another you guys say, 'Ah, now we've got two gaps where there, where previously there was only one.' I mean, almost every fossil you find is intermediate between something and something else."[6]

These snapshots of species from the fossil record allow scientists to infer how evolution transpired, but the fossil record itself will always be incomplete, although every new discovery makes it less so. Recent discoveries have substantially enriched the hominin fossil record, but some, perhaps many, hominin species are without doubt missing from it. Some left no fossils because of where they lived and how they died. Others probably left fossils, but we have yet to find them.

The image of evolution as a chain, with each species as a link in the chain, is misleading. The chain model—better referred to as a *lineage*—is correct in the sense that there is an unbroken lineage leading from one ancestor to another. For instance, you can trace your maternal lineage from your mother, to her mother, then to her mother, and so on. However, tracing a single lineage disregards all the other lineages, such as those leading to distant relatives. Relatives not in your direct line of ancestry, such as an aunt or cousin, can offer much genetic information about your extended family, even though they are not your ancestors. Likewise, in evolutionary science, members of related lineages offer useful information about our ancestors even if they are not in our direct line of ancestry.

A better representation of hominin evolution is a branching tree: the branch points represent divergences of lineages from common ancestry, and the branches represent the lineages themselves. Charles Darwin was apparently the first to depict evolution as a tree in a sketch he drew in 1837 (figure 3.4a). Especially appealing are the words *I think* written above it. Representation of evolution as a tree has the added advantage of being an easy way to depict lineages that go extinct, which Darwin showed in his more refined evolutionary tree, published as his only drawing in *The Origin of Species* (figure 3.4b).

Currently, there is only one certain way to determine whether an extinct species is ancestral to one now living, and that is to conduct a large-scale comparison of the DNA, preferably of the entire genomes, of the extinct and living species. As we'll see shortly, this is now possible for some of our most recent hominin relatives, such as Neanderthals and Denisovans. However, the DNA that once was present in older fossil hominins is now entirely degraded,

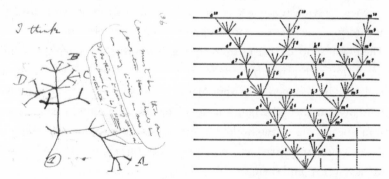

a. Drawing from Darwin's notebook.   **b.** Drawing from *Origin of Species*.

Figure 3.4. Darwin's drawings of evolutionary trees.

so there is currently no method scientists can use to determine with absolute certainty whether or not these older fossil species were ancestral to humans or whether they are from side branches in the hominin family tree that went extinct. Given the large number of species that existed in the past, we can safely conclude that many hominin fossils are probably *not* the direct ancestors of humans but instead represent nearby side branches.

Even if fossilized hominin species are not ancestral to a modern species, they nonetheless tell us much about the actual ancestral species because they were closely related to our ancestors, and in them we see how and approximately when certain uniquely human anatomical features evolved.

A SEVEN-MILLION-YEAR JOURNEY THROUGH HOMININ EVOLUTION

Let's now travel through the fossil evidence of our extinct hominin relatives, beginning about seven million years ago and ending with the present. The hominin fossils discovered so far are pieces of an incomplete puzzle, and new pieces are being discovered at an increasing pace. Enough pieces are present to show us a broad and deep evolutionary picture of our distant humanlike relatives, and from them we can derive three key points. First, all of the oldest hominin fossils, covering a period from about seven million until two million years ago, are found exclusively in Africa. The earliest hominin fossils discovered outside Africa date to slightly less than two million years ago, and all of them are members of our own genus, *Homo*, with anatomical features similar to ours. Second, certain traits, such as upright posture and bipedalism evolved early, whereas increased brain size and tool usage arose later. And the transitions from earlier to later forms are readily evident. Third, the oldest fossils of our own species, *Homo sapiens*, are also from Africa—strong evidence that modern humans originated in Africa even though earlier *Homo* species had already established themselves elsewhere in the world. Information from DNA analysis in modern humans confirms that modern humans originated in Africa, as we'll see in the next chapter.

As shown in table 3.1, hominin fossils fall into four major groups based

TABLE 3.1. FOUR GROUPS OF HOMININ SPECIES

Species	Approximate Time Span (mya = millions of years ago)	Geographic Distribution
Group 1: Early Hominins		
Sahelanthropus tchadensis	~7 mya	Africa: Chad
Orrorin tugenensis	6.2–5.65 mya	Africa: Kenya
Ardipithecus kadabba	5.77–5.18 mya	Africa: Ethiopia
Ardipithecus ramidus	4.42–3.9 mya	Africa: Ethiopia
Group 2: Australopithecines		
Australopithecus anamensis	4.17–3.95 mya	Africa: Kenya
Kenyanthropus platyops	3.5–3.3 mya	Africa: Kenya
Australopithecus afarensis	3.76–2.92 mya	Africa: Ethiopia, Kenya, Tanzania
Australopithecus africanus	4–2.5 mya	Africa: South Africa
Australopithecus garhi	2.46 mya	Africa: Ethiopia
Australopithecus sediba	1.95–1.78 mya	Africa: South Africa
Group 3: Robusts		
Paranthropus aethiopicus	2.8–2.3 mya	Africa: Ethiopia
Paranthropus robustus/crassidens	2–1 mya	Africa: South Africa
Paranthropus boisei	2.3–1.4 mya	Africa: Ethiopia
Group 4: Homo		
Homo rudolfensis	1.9 mya	Africa: Kenya
Homo habilis	1.83–1.53 mya	Africa: Kenya, Tanzania
Homo erectus (ergaster)	1.9–1.49 mya	Africa: Ethiopia, Kenya, South Africa
Homo erectus (georgicus)	1.8 mya	Asia: Georgia
Homo erectus	1.9–0.03 mya	Southeast Asia: Indonesia (Java)
Homo pekinensis	0.6 mya	Asia: China
Homo antecessor	0.8 mya	Europe: Spain (possibly Africa: Algeria)
Homo rhodesiensis	0.7–0.2 mya	Africa: Ethiopia, Rhodesia, South Africa
Homo heidelbergensis	0.7–0.2 mya	Africa: Morocco
		Europe: Spain, France, England, Germany, Hungary, Greece
		Middle East: Israel
Homo neanderthalensis	0.2–0.03 mya	Europe: Portugal, Spain, France, Belgium, Germany, Czech Republic, Italy, Croatia, Hungary
		Middle East and West Asia: Georgia, Israel, Iraq, Uzbekistan
Homo sp. (Denisovan)	0.5–0.2 mya	Asia: Russia (Siberia)
Homo floresiensis	0.07–0.01 mya	Southeast Asia: Indonesia (Flores)
Homo sapiens	0.2 mya–present	Worldwide

Dates and geographic distributions are from G. L. Sawyer et al., *The Last Human: A Guide to Twenty-Two Species of Extinct Humans* (New Haven, CT: Yale University Press, 2007), except for *Australopithecus sediba*, which is from L. R. Beger et al., "*Australopithecus sediba*: A New Species of *Homo*-Like Australopith from South Africa," *Science* 328 (2010): 195–204, and *Homo sp.* (Denisovan), which is from D. Reich et al., "Genetic History of an Archaic Hominin Group from Denisova Cave in Siberia," *Nature* 468 (2010): 1053–60.

on when they lived and their anatomical features. We won't take the time here to discuss the details of each species but will instead focus on a few of the most complete and best-preserved examples.[7]

The first and oldest group consists of the early hominins, which date from about 7 to 4.4 million years ago and are all from Africa. The teeth of all known members of this group are similar, which has led some of the foremost experts to suggest that they should be grouped in the same genus.[8] The second group includes the australopithecines, which are well represented in the fossil record. All lived in Africa from about 4 million to 2.5 million years ago. The third and fourth groups are the robusts and the genus *Homo*, the latter including early and modern humans. These two groups probably diverged from common australopithecine ancestry about 2 million years ago in Africa. The robust branch declined in Africa and suffered extinction, whereas early humans spread throughout Africa and several times ventured beyond Africa to colonize other parts of the world.

THE EARLY HOMININS

Figure 3.5 (page 78) depicts the locations where remains from the three species of early hominins have been found. The oldest known hominin to date, *Sahelanthropus tchadensis*, is represented by a nearly complete cranium from an individual nicknamed Toumai, meaning "hope of life." This cranium displays some of the oldest evidence of upright posture. As we saw in the previous chapter, at the base of vertebrate skulls is an opening called the foramen magnum, where the brain stem protrudes out of the skull into the vertebral column. In humans, the foramen magnum's position is underneath the cranium, where it points downward, a position that is better suited for habitual upright posture. In chimpanzees and other great apes, it is toward the back of the cranium and points backward, consistent with their forward-leaning posture.

As figure 3.6 (page 79) shows, Toumai's cranium is very apelike, with a protruding jaw, a prominent brow, and a nearly horizontal forehead. But the foramen magnum is situated between the human and chimpanzee positions

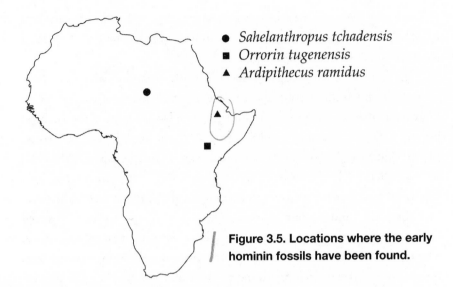

- *Sahelanthropus tchadensis*
- *Orrorin tugenensis*
- *Ardipithecus ramidus*

Figure 3.5. Locations where the early hominin fossils have been found.

and points downward, suggesting that this ancient species probably had at least a partially upright posture. Unfortunately, no skeletal remains beyond the skull of Toumai's species have been reported.

The latest species in this oldest group of early hominins is *Ardipithecus ramidus*, dating to 4.4 million years ago. The fossil record is exceptionally rich for this species and includes a partial skeleton of a female nicknamed "Ardi," for *Ardipithecus*. The October 2, 2009, issue of the journal *Science* features Ardi on its cover and contains a series of articles analyzing Ardi's skeleton and the remains of additional individuals of her species, as well as fossils from other animal and plant species found near her. The news made huge fanfare in scientific circles and in the popular press because Ardi's skeleton was so complete and was such an important find.[9]

Our most recent common ancestor with chimpanzees lived sometime between five and seven million years ago, and Ardi's age places her closer in time to that ancestor than to us. So we might expect her to have many apelike characteristics, and indeed she does. But other characteristics are not consistent with those of modern apes. The scientists who reconstructed her skull wrote that it has "a decidedly ape-like gestalt" yet is "not particularly chimpanzee-like,"[10] as is evident in figure 3.7 (page 80). Her arms, hands, legs, and feet do not have the knuckle-walking adaptations of modern great

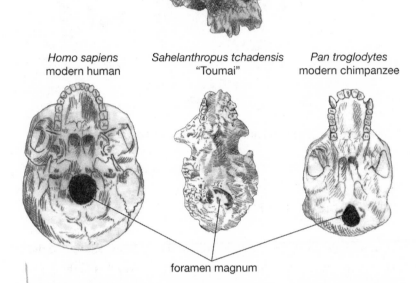

Homo sapiens
modern human

Sahelanthropus tchadensis
"Toumai"

Pan troglodytes
modern chimpanzee

foramen magnum

**Figure 3.6. Positions of the foramen magnum in modern humans, in
Sahelanthropus tchadensis (Toumai), and in modern chimpanzees.**

apes nor the characteristics for efficient upright walking and running. Instead, they portray a species well adapted for climbing in trees.[11]

One of Ardi's most remarkable features is her pelvis. The ilia (the upper hipbones in the pelvis, which we feel when we place our hands on our hips) are very much like a human's, but her lower pelvis more closely resembles an ape's (figure 3.7, page 80). Her pelvis (and other features) indicates that she was a *facultative biped*—a species adapted for walking on two feet when on the ground but not a habitual walker. According to the scientists who analyzed and reconstructed Ardi's pelvis: "Although the foot anatomy of *Ar. ramidus* showed that it was still climbing trees, on the ground it walked upright."[12] Upright bipedalism was not yet habitual, but it had already evolved in hominins by 4.4 million years ago.

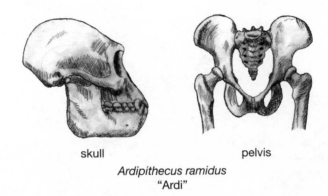

skull   pelvis

Ardipithecus ramidus
"Ardi"

Figure 3.7. The skull and pelvis of *Ardipithecus ramidus* ("Ardi").

THE AUSTRALOPITHECINES

Let's move forward about a million years to Lucy, as an example of the second major group, the australopithecines, which, by the way, has nothing to do with Australia—*austral* means "southern" and *pithecine* means "ape." Our own genus, *Homo*, almost certainly evolved from australopithecines, so their fossils offer us compelling clues about our ancient ancestry. Australopithecine fossils are all from Africa and date from about 4 million to 2.5 million years ago.

Lucy is one of the most famed fossils ever unearthed, a partial skeleton of a female who lived about 3.4 million years ago in what is now Ethiopia, about a million years after Ardi. Also representing Lucy's species are the partial remains of thirteen individuals, nicknamed First Family, who died together in the same place, and the nearly complete skull and partial skeleton of a three-year-old female nicknamed Selam, which means "peace." Because she was a young child, Selam is also sometimes called Lucy's baby, although she lived about 120,000 years after Lucy. The *afarensis* in their Latin name *Australopithecus afarensis* means they lived in the Afar region of Africa.

The *pithecus* in Lucy's Latin name means "ape," and Lucy's skull retains many apelike characteristics, with a protruding mandible and jaw, a prominent brow, no nose bridge, very little forehead, and a proportionally small

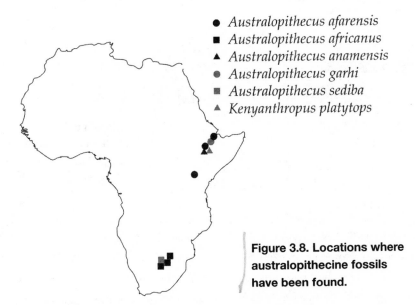

● *Australopithecus afarensis*
■ *Australopithecus africanus*
▲ *Australopithecus anamensis*
● *Australopithecus garhi*
■ *Australopithecus sediba*
▲ *Kenyanthropus platytops*

Figure 3.8. Locations where australopithecine fossils have been found.

brain compared to modern humans. By contrast, much of the rest of her body was more humanlike than apelike. Her pelvis closely resembles a human pelvis in both its upper and lower portions (figure 3.9, page 82). Her other anatomical features—such as arms, legs, hands, and feet—show that she was quite capable of walking upright on the ground, although she was not as well adapted to bipedalism as we are. Her anatomy also reveals a species that was not as adept at tree climbing as Ardi but was much better than are modern humans. Much of Lucy's anatomy represents a transition between an arboreal (tree climbing) ape and an upright-walking human.

Two spectacular lines of evidence conclusively tell us that Lucy's species was habitually bipedal. The first is footprints made in wet volcanic ash, which then solidified, at two sites near the town of Laetoli in Tanzania. One site has footprints from a single individual, and the second site has side-by-side sets of footprints by two individuals who were probably adults, and possibly a third set from a smaller individual, perhaps a child, who may have stepped in the footprints of one of the others.[13] The footprints are clearly bipedal with no evidence of forelimbs touching the ground. Modern apes sometimes walk on two feet, however. This fact led some scientists to question whether the australopiths who made these footprints were truly bipedal or if they were quadrupeds

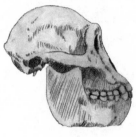
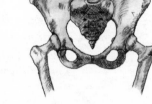

skull   pelvis

Ausralopithecus afarensis
"Lucy," "First Family," "Selam"

Figure 3.9. The skull and pelvis of *Australopithecus afarensis*.

choosing to walk bipedally across sandy-wet volcanic ash as modern great apes sometimes do. However, an extensive analysis of the footprints shows that the individuals who made them walked fluidly and efficiently, using a gait similar to that of modern humans rather than the awkward gait of modern apes that occasionally walk on two feet.[14]

The second line of evidence is a metatarsal (foot) bone, recently discovered and published in the journal *Science* in 2011.[15] Although Lucy's skeleton is fairly complete, missing from it, and from the remnants of other members of her species, were key bones in the feet. The arched shape of this newly discovered metatarsal bone is very similar to the corresponding bone in modern humans and conclusively shows that Lucy's species was habitually bipedal.

The first evidence of tool use comes from one of the latest australopithecines, *Australopithecus garhi*, which lived about 2.5 million years ago. Although the tools themselves have not been found, broken antelope bones that were apparently broken open with stone tools to reach the nutrient-rich marrow were found with remains of this species.[16]

We cannot be certain when tool use first evolved. Modern chimpanzees and orangutans (as well as some modern bird species) use sticks as rudimentary tools to access insects for food, and they may fashion the sticks to make them more efficient. Tools such as sticks, twigs, and plant fibers used for baskets or slings consist of materials that rapidly decay and are less likely to persist as evidence of tool use. Stone tools, however, persist well. Although no stone tools attributable to *Australopithecus garhi* have yet been discovered, stone tools are common with remains of members of the genus *Homo*, and some with *Paranthropus*, as we'll see shortly.

THE ROBUSTS AND EARLY HUMANS

Between three and two million years ago, two major branches diverged from the australopithecines in the hominin family tree—the robusts and the early humans—and they constitute the third and fourth major groups of hominins. The name *robust* means large and strong, and the robusts had features of a powerful body and bite. One of their most distinctive features is a crest running front-to-back along the tops of their skulls, where massive chewing muscles attached.

Some scientists classify robusts as *Australopithecus* (Lucy's genus) because they share many characteristics in common with the australopithecines. Others argue that they should be in a separate genus, *Paranthropus*, because of their distinct anatomical characteristics.

There is some evidence of tool use by robusts dating back as far as 2.4 million years ago. This evidence consists of bone tools showing wear marks, burns on bone tools consistent with fire use, and stone tools that had been shaped. Some of the tools are found in deposits that contain remains of both *Paranthropus* and *Homo*, but others are in sites that contain only *Paranthropus* remains, suggesting that species of *Paranthropus* used tools and possibly fire.[17]

The robusts were not our ancestors but instead comprise a side branch. Some of them lived in the same areas and at the same times as early humans in Africa, but they went extinct as early humans were spreading and diversifying throughout and beyond Africa. Fossils of the robusts date from about 2.8 to 1.2 million years ago, and all are from Africa.

Figure 3.10. A robust cranium. Notice the crest along the top to which strong chewing muscles attached.

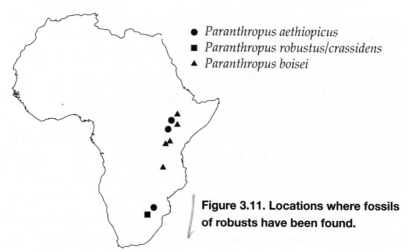

Figure 3.11. Locations where fossils of robusts have been found.

The fourth group is the human group, and its members are classified in the genus *Homo*. This group has a long history of evolution in Africa from about two million years ago until the present. As members of this genus evolved, they began spreading and diversifying throughout Africa. Their fossils are found in Africa as far north as the Mediterranean coast to near the continent's southern tip. The three oldest *Homo* species discovered so far, *Homo rudolfensis*, *Homo habilis*, and *Homo erectus* (*ergaster*), date to between 1.9 and 1.5 million years ago, and all are African.

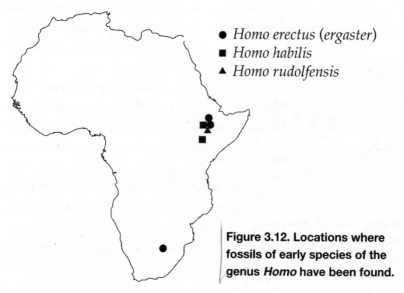

Figure 3.12. Locations where fossils of early species of the genus *Homo* have been found.

There is considerable debate about how these early humans should be classified. For example, some scientists argue that a species discovered in 2010 in South Africa and named *Australopithecus sediba* should in fact be classified as *Homo*. Even though its discoverers named it *Australopithecus*, they referred to it as "*Homo*-like."[18] Because the anatomical features of these early *Homo* species are transitional between australopithecines and later species of *Homo*, arguments abound over whether they should be classified as *Australopithecus* or as *Homo*. Regardless of their assigned classification, their remains offer evidence that the transition from australopithecines to early humans was gradual rather than sudden.

By the time members of the genus *Homo* emerged, about a million years after Lucy, stone tool use was common. Stone tools are associated with most species of *Homo*, beginning with early *Homo* species in Africa and continuing with advances in tool use among later species of *Homo* in Africa and those that migrated beyond Africa. An early African species, *Homo habilis*, in fact, is named because of the tools found associated with its remains; the name *habilis* means "handyman." In particular, stone tools that have been shaped on two sides to form a sharp edge are present with the remains of early *Homo* in Africa.[19]

Also, the emergence of *Homo* coincides with abundant evidence throughout the entire skeleton of habitual bipedalism. A species whose early African members may have been ancestral to us is *Homo erectus*. As we'll discuss in some detail momentarily, this species ventured beyond Africa, and its African specimens are sometimes called *Homo ergaster*, or *Homo erectus* (*ergaster*), to distinguish them from the non-African specimens of *Homo erectus*.

One of the most spectacular finds of *Homo erectus* (*ergaster*) is a nearly complete skeleton nicknamed Turkana Boy (also called Nariokotome Boy) because his remains were discovered in the Nariokotome region near Lake Turkana in Kenya. He died when he was between nine and twelve years old about 1.5 million years ago. His skull has many humanlike features, including vertically set teeth, a larger braincase when compared to earlier hominins, and evidence of a bony protrusion below the nose bridge, characteristic of a protruding, humanlike nose. Like australopithecines, however, his brow is prominent across the face, and his forehead slopes back (figure 3.13, page 86). Below the skull, however, his skeleton is very much like a modern human's,

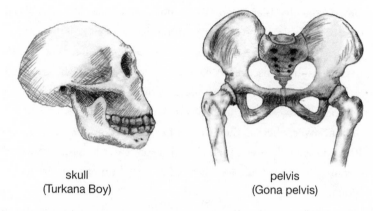

skull
(Turkana Boy)

pelvis
(Gona pelvis)

Homo erectus (ergaster)

Figure 3.13. The skull and pelvis of *Homo erectus* (*ergaster*), the African type of *Homo erectus*.

with all the adaptations for complete, habitual upright walking and running. Although his head has many characteristics of an australopithecine, the rest of his anatomy is very much like ours.

A recent African *Homo erectus* discovery is the pelvis of an adult female, dating from 0.9 to 1.4 million years ago, found in Gona, in the Afar region of Ethiopia.[20] This is the oldest hominin pelvis discovered with a wider opening, similar to that of modern humans, to accommodate a relatively large brain size during childbirth (figure 3.13, above). This pelvic opening indicates that by one million years ago larger brains had evolved in our ancestors.

Three of the species we've just discussed, *Ardipithecus ramidus* (Ardi), *Australopithecus afarensis* (Lucy, First Family, and Selam), and *Homo erectus* (Turkana Boy and the Gona pelvis), are well-preserved examples of hominin evolution, each spaced about a million years apart. For a moment, let's look at all of them together (figure 3.14, page 87), comparing them to modern humans to see how the skull and the pelvis have evolved over the past 4.4 million years. Such a comparison shows that many of the features in the modern human pelvis, especially those favoring upright posture and bipedalism, evolved early, whereas the features of the human skull, especially the increase in brain size and widening of the pelvis to accommodate it, appeared later.

Ardipithecus ramidus
skull

Ardipithecus ramidus
pelvis

Ausralopithecus afarensis
skull

Ausralopithecus afarensis
pelvis

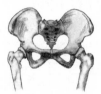

Homo erectus (ergaster)
skull

Homo erectus (ergaster)
pelvis

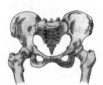

Modern Chimpanzee
Pan troglodytes
skull

Modern Human
Homo sapiens
skull

Modern Human
Homo sapiens
pelvis

Modern Chimpanzee
Pan troglodytes
pelvis

**Figure 3.14. Comparison of the skulls and pelvises of three
hominins with those of modern humans and chimpanzees.
All pelvises are from females.**

Even so, the earliest hominins had clearly diverged in some aspects from their ancestral apelike anatomy.

HOMININS MIGRATED OUT OF AFRICA AT LEAST THREE TIMES.

According to fossils discovered so far, all hominins apparently remained in Africa until the emergence of *Homo* about two million years ago. Since that time, several groups ventured beyond Africa with at least three sets of migrations, each separated from the previous migration by hundreds of thousands of years. The two earlier migrations resulted in the evolution of new species or subspecies beyond the confines of Africa. The third did not result in any new species but instead founded a series of sequential migrations over a period of sixty thousand years when modern humans spread throughout the world.

Let's begin with the oldest migration. Members of *Homo erectus* (*ergaster*) left Africa and migrated into the Arabian Peninsula around 1.9 million years ago. As depicted in figure 3.15, below, their descendants then migrated

Figure 3.15. Fossil locations and probable migrations of *Homo erectus*.

across the Middle East and Asia, and some to Southeast Asia. Evidence of their migration and settlement comes from their remains discovered in central Asia, China, and Indonesia.

Some of the oldest fossils from this group, dating to about 1.8 million years ago, are from the town of Dmanisi in the nation of Georgia between the Black and Caspian Seas. These are classified as Homo erectus (georgicus). Also dating to about 1.8 million years ago are fossils of Homo erectus on the Island of Java in Indonesia. Descendants of these early immigrants remained in Java for a very long time—until 120,000 years ago, and perhaps as recently as 27,000 years ago. Over time they changed enough that some scientists contend that we should classify them into at least three species: Homo modjokertensis, Homo erectus, and Homo soloensis.

A branch known as Homo erectus (pekinensis) migrated eastward into what is now China, represented by fossils found in a cave near Beijing. These remains date to about six hundred thousand years ago. Another possible branch is from the Island of Flores in Indonesia where a small-statured species named Homo floresiensis lived as recently as 12,000 years ago. Adults were only about a meter tall, and evidence suggests that they represent a lineage that diverged from the Homo erectus lineage as early as several hundred thousand years ago. Some scientists have argued, however, that they instead were modern humans with a medical disorder that caused their shortened stature. No doubt the debate on their origin will continue.[21] Efforts are under way to hopefully obtain DNA from their remains, which could definitively answer the question of whether they were modern humans or a separate species.

The second migration out of Africa took place across the Straits of Gibraltar between what are now Morocco and Spain, beginning about 800,000 years ago (figure 3.16, page 90). The African species from which these migrants descended is unknown, although Homo rhodesiensis lived at the right time and is a plausible candidate. The oldest fossils found in Spain are from a species named Homo antecessor, which dates to about 780,000 years ago. Fossils of a later, closely related species, named Homo heidelbergensis, are widespread throughout Europe and the Middle East, found in Morocco, Spain, France, England, Germany, Hungary, Italy, Greece, and Israel, dating from about 700,000 until 200,000 years ago.

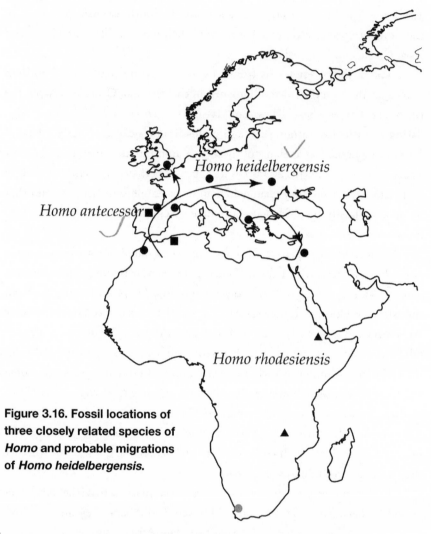

Figure 3.16. Fossil locations of three closely related species of *Homo* and probable migrations of *Homo heidelbergensis*.

Later specimens of *Homo heidelbergensis* so strongly resemble early specimens of Neanderthals (*Homo neanderthalensis*) that it is difficult to draw a line between the two, an observation offering strong evidence that Neanderthals evolved in Europe from *Homo heidelbergensis*. The oldest skeletons classified as Neanderthals are from France and date to about 175,000 years ago.[22]

Neanderthals are perhaps the most famous of all extinct hominin species and have yielded the most abundant remains. Some skeletons are nearly complete, and the combined information from these skeletons has revealed much about their anatomy. We also know quite a lot about their tools, clothing, communities, and culture from the remains they left behind. In many respects, they are similar to modern humans, which led some anthropologists to classify them as a subspecies of humans (*Homo sapiens neanderthalensis*). Not long ago, people speculated that modern humans descended from Neanderthals in various places. However, an abundance of evidence—from anatomy, geography, and DNA—supports a very different scenario.

According to this scenario, Neanderthals evolved from *Homo heidelbergensis*, most probably in Western Europe about 200,000 to 175,000 years ago. They spread throughout Europe and into the Middle East and west-central Asia as far east as Uzbekistan (figure 3.17). They lived until about 30,000 years ago, when the last populations dwindled and went extinct in southern Europe along the Mediterranean coast.[23]

Our ancestors remained in Africa where the earliest fossils with clear modern human anatomy emerged about two hundred thousand years ago, which marks the beginning of our own species—*Homo sapiens*. The majority of

Figure 3.17. Fossil locations and probable migrations of *Homo neanderthalensis*.

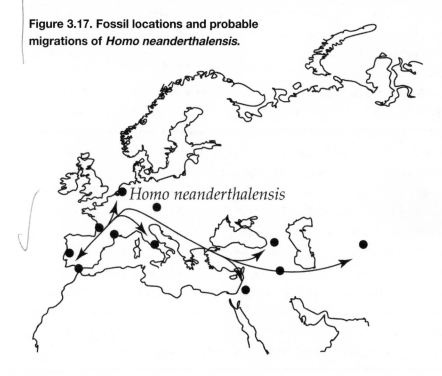

Homo neanderthalensis

these early people remained in Africa and are the ancestors of modern native Africans. About sixty thousand years ago, a relatively small group of people left Africa, migrating into the Arabian Peninsula, and they became the ancestral founders of people who through many generations spread to colonize the rest of the world. Their descendants migrated northward into the Middle East and west-central Asia, which became a staging ground for further waves of migration, ultimately extending to the far reaches of the world. Several groups migrated west into Europe and the Middle East, settling in the regions where Neanderthals lived, with whom they coexisted.

How did modern humans manage to spread throughout the earth, increasing in numbers at the same time that Neanderthals dwindled and ultimately went extinct? Much speculation exists about the role humans might have played in the Neanderthal demise. Did humans drive Neanderthals to extinction through competition and possibly genocide,[24] were Neanderthals assimilated into human populations through interbreeding, or did Neanderthals go extinct from some other assault that humans were better able to endure, such as disease or climate change? Although we don't have complete answers to these questions, some important clues led scientists to the shocking conclusion that possibly everyone alive today whose ancestry is from outside Africa carries some Neanderthal DNA.

A team led by evolutionary biologist Svante Pääbo from the Max Planck Institute in Leipzig, Germany, successfully sequenced Neanderthal DNA and compared it with DNA variations known to exist in humans. The team's evidence shows that people of pure African ancestry share no DNA with Neanderthals. This is no surprise because there is no evidence that Neanderthals lived in Africa. But people whose ancestry is outside of Africa probably inherited about 1 to 4 percent of their DNA from Neanderthals. This DNA probably came from limited mating between humans and Neanderthals, but the proportion of Neanderthal DNA is so small that humans quite certainly did not assimilate Neanderthals through widespread mating. Moreover, although Neanderthals occupied regions extending from Europe to west-central Asia, people from east Asia, where Neanderthals were apparently not present, likewise carry Neanderthal DNA. This evidence suggests that the human-Neanderthal mating took place early, about the time when humans first encountered Neanderthals in the Middle East soon after migrating out of Africa.[25]

Analysis of ancient DNA shows that Neanderthals were not the only hominin species encountered by humans as they spread throughout the world. In 2008, archaeologists excavated a hominin finger bone, dating to between fifty thousand and thirty thousand years ago, in the Denisova Cave in southern Siberia. Pääbo's team extracted DNA from the bone's interior and compared it to DNA from modern humans and Neanderthals. The Denisovan DNA is different enough from the DNA of both humans and Neanderthals to indicate that it belongs to a separate species—and this species is probably an offshoot from early Neanderthals. Moreover, Pääbo's team found evidence that this species may have interbred with a group of humans who were the ancestors of modern Melanesians. According to the most likely scenario, the ancient ancestors of Melanesians migrated through central Asia, where some of them mated with Denisovans. Thereafter, their descendants continued to migrate to Southeast Asia and ultimately into the Pacific Islands, where their modern descendants now live.[26]

Most of what we know about the first two hominin migrations out of Africa comes from fossil evidence and geography, with some recent and fascinating information about Neanderthals and Denisovans from their DNA. Our own species has lived and migrated throughout Africa since its origin about two hundred thousand years ago, and it also migrated out of Africa to colonize the rest of the world beginning about sixty thousand years ago. Archaeological evidence tells us much about their histories and migrations both in and out of Africa. But our understanding of them has exploded in recent years, thanks largely to evidence from DNA. We explore that evidence and what it tells us in the next chapter.

CHAPTER 4
EVIDENCE FROM GEOGRAPHY

According to ancient Roman mythology, newborn twin brothers Romulus and Remus were cast by their jealous uncle into the Tiber River to die. A female wolf rescued and suckled the two infants, and a shepherd later raised the boys to adulthood. Eventually, the two brothers returned to the Capitoline Hill, where the wolf had nourished them, and there they founded a city. Romulus, in a fit of rage after Remus had mocked him, killed his brother and became the sole leader of the new city that henceforth bore a name derived from his—Rome. A famous bronze statue of the she-wolf suckling the two infants now stands in Rome at the Capitoline Museum.

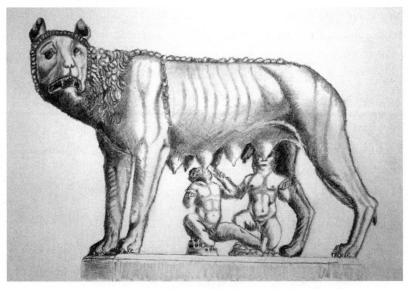

Figure 4.1. A she-wolf suckling Romulus and Remus.

95

The story is purely mythical, but the powerful empire that emerged from Rome is well-chronicled history. Although the Romans left a vast network of roads, aqueducts, buildings, and some of the world's most exquisite art, perhaps their most enduring legacy is their language. Throughout the continent and islands of western Europe in what is now Italy, France, Spain, and Portugal, and in a region of eastern Europe in what is now Romania, more than sixty related languages and dialects emerged from Latin, the language of ancient Rome.[1] These languages spread to the Americas and Africa and to more limited parts of the rest of the world. Even English, although technically not a Romance language (the branch of languages that descended from ancient Rome), acquired a substantial portion of its vocabulary and some of its grammatical constructions from Latin through old French.

The five major Romance languages—Italian, Spanish, Portuguese, French, and Romanian—as well as a host of other languages, have diverged to the point that a native speaker of one has some difficulties fully understanding another without studying it. However, enough similarities persist—in vocabulary, pronunciation, syntax, and grammar—that speakers of one Romance language are usually able to perceive the general meaning when they hear another Romance language spoken or when they read it. For example, my favorite of Michelangelo's sonnets contains the verse (in English translation):

> *I fly without feathers upon your wings;*
> *By your genius I am forever carried to the sky.*[2]

The following are renderings of this same verse, first in Michelangelo's original Italian, then in Spanish, Portuguese, French, and Romanian:

> *Volo con le vostr'ale senza piume;*
> *Col vostro ingegno al ciel sempre son mosso.*

> *Vuelo con vuestras alas sin plumas;*
> *Con vuestro genio al cielo siempre soy movido.*

> *Vôo com vossas asas sem plumas;*
> *Com vosso gênio ao céu sempre sou movido.*

Je vole avec votr'ailes sans plumes;
Avec votre génie au ciel toujours je suis ému.

Zbor cu aripile voastre fără pene;
Cu geniul vostru spre ceruri mereu eu sunt minat.[3]

The similarities are readily apparent and explained by their lingual derivation from a common root language. For example, the English word *genius* is derived (through Old French) from the Latin word *ingenium*. It became *ingegno* in Michelangelo's Italian, *genio* in Spanish, *gênio* in Portuguese, *génie* in French, and *geniul* in Romanian.

Linguists have successfully reconstructed the history of the Romance languages by comparing the similarities and differences in modern languages and dialects with ancient versions preserved in old manuscripts. These linguistic comparisons naturally overlay on the geography and settlements of what was the Roman Empire.

The parallels between the evolution of language and the evolution of life are striking, and for good reason: like DNA, language is an imperfect replicator. Each person learns language though imitation during childhood from those who surround her or him. The child expertly mimics the sounds, word constructions, and meanings learned from others, a consequence of a brain genetically programmed by evolution to coordinate and understand speech. Over a relatively short period of time, the child attains native fluency, and in so doing, replicates the language into the next generation. That child may pass on the language to siblings, friends, and, eventually, to her or his children, who replicate it in yet another generation.

Languages evolve because the replication is imperfect. A mispronunciation repeated by one person in one generation may become an acceptable pronunciation for several people in the next. Over time, modified words, or grammatical constructions, may spread, eventually attaining wide use in an emerging dialect and ultimately in a language. For much of human history, modified pronunciations replaced older ones, and the same is true of grammatical constructions. Through preserved written materials, scholars can observe and reconstruct much of their evolution. More recently, through electronic recordings, we can hear changes emerging over just a few generations. Listen to a television show or movie made forty years ago, and some of the changes in pronunciation and word use in our current languages are obvious.

There is, however, a major difference between language evolution and biological evolution. Although the human ability to learn a language (or even several languages) is inherited—encoded within our DNA—*which* language a person learns is not. A child learns the language spoken in her or his environment, which may or may not be the language of her or his biological parents. Variations in language are replicated from one generation to the next as learned behaviors, not as changes in our DNA. By contrast, biologically inherited variations *are* encoded in our DNA and are transmitted from parents to offspring through biological reproduction. Evolution, at its most basic level, is absolutely and inextricably dependent on inherited variations in DNA. And the ultimate story of evolution is written into the DNA of every person, and every organism, who now lives or ever lived.

Only in recent decades have scientists been able to read evolutionary history in DNA. The ability to do so dates back to the 1970s, when Frederick Sanger developed a method of DNA sequencing that has since become automated. In 1977, he published the first complete DNA sequence of an organism— the DNA sequence belonging to the bacterial virus ΦX174.[4] This feat, which took him months to accomplish, can now be completed in a few hours. Recent advances in DNA sequencing technology have resulted in automated DNA sequencers capable of high-throughput sequencing, a very rapid form of sequencing that analyzes enormous amounts of DNA sequence in a short period of time. The sequencers are commonplace in today's laboratories. And thousands of scientists everywhere routinely submit the DNA sequences they discover to centralized databases, freely available to all scientists everywhere over the Internet—a vast and open worldwide collaboration on a scale unprecedented in the history of science. The result is a massive resource of information that increases with each passing day.

From these DNA sequences, countless histories of evolution are emerging. Some of the most fascinating are the ancient human diasporas—where and when our species originated, and how ancient people populated the world. Like linguists who overlay language development on geography, evolutionary biologists are integrating information from DNA with geography to reconstruct how people migrated in multiple waves from the region where our species originated in Africa to ultimately inhabit the world.

HOW DOES DNA REVEAL ANCIENT HUMAN HISTORY?

To understand how scientists read evolutionary history in DNA, we need to review a few key features about DNA. First, the DNA molecule has a beautiful double-helical structure, two linear strands wound around each other with steps holding them together, reminiscent of a spiral staircase. Each strand contains a digital code that scientists represent with four letters, T, C, A, and G. These letters are called *bases* to denote their chemical properties.

Second, scientists write DNA sequences using the base letters T, C, A, and G just as we use an alphabet to write words. Although DNA is a double helix, we can mentally unwind the sequence and read it as a written line of bases. For example, the following sequence is a small segment of the DNA in the human insulin gene:

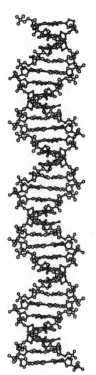

Figure 4.2. The DNA molecule.

GGCCCTCTGGGGACCTGACCCAGCCGCAGCCTTTGTGAAC

Third, because DNA molecules are double-stranded, each strand in DNA—each line in the written sequence—has a partner strand. The bases in the two partner strands are paired so that every A in one strand is paired with a T in the other strand, and every G is paired with C. Thus, we can write the double-stranded version of that insulin-gene segment we just examined as:

GGCCCTCTGGGGACCTGACCCAGCCGCAGCCTTTGTGAAC
CCGGGAGACCCCTGGACTGGGTCGGCGTCGGAAACACTTG

Fourth, this pattern of base pairing makes replication of the DNA molecule possible, and replication is one of the features that is absolutely essential for the inherited material. The two strands separate, and specialized proteins assemble new strands paired with the old ones by adding new bases using the T-to-C, A-to-G base-pairing rules. The old strands serve as templates to define the sequence of the new strands. The result is two identical, double-stranded DNA molecules derived from one original DNA molecule:

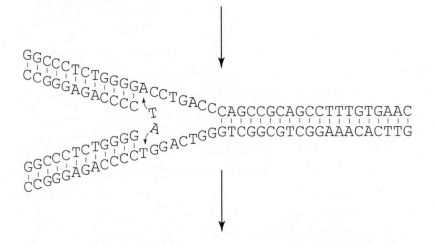

Lastly, DNA sequences may mutate. Rarely, a mistake during DNA replication—sometimes purely an error, other times the result of exposure to radiation or certain chemicals—may change the sequence, such as a T-A pair replaced by a C-G pair:

```
GGCCCTCTGGGGACCTGACCCAGCCGCAGCCTTTGTGAAC
 | | | | | | | | | | | | | | | | | | | | | | | | | | | | | | | | | | | | | | | |
CCGGGAGACCCCTGGACTGGGTCGGCGTCGGAAACACTTG
```

$$\downarrow$$

```
GGCCCTCTGGGGACCTGACCCAGCCGCAGCCCTTGTGAAC
 | | | | | | | | | | | | | | | | | | | | | | | | | | | | | | | | | | | | | | | |
CCGGGAGACCCCTGGACTGGGTCGGCGTCGGGAACACTTG
```

Once the mutation is established in both strands of the DNA, it is faithfully replicated from that time forward. The particular mutation illustrated in this example has occurred in the human insulin gene and causes a rare form of inherited childhood diabetes.

MITOCHONDRIAL AND Y-CHROMOSOME DNA SEQUENCES REVEAL DISTINCT PATTERNS OF HUMAN EVOLUTION.

Having reviewed these basic aspects of DNA, we're ready to predict what we should see in DNA as a consequence of evolution. If we were to compare the complete sequence of my DNA with yours, our two sequences would be almost identical—in fact, extremely close to identical (excluding the Y chromosome for female-male comparisons). Any two people selected at random from the earth's population are more than 99 percent identical throughout most of their DNA sequences.

Certain types of DNA, however, tend to vary more than the rest. One type is the DNA in our mitochondria, which are small, bacteria-like structures in our cells, much like tiny cells within our cells. The vast majority of our DNA sequence resides in the cell nucleus and is called *nuclear DNA*, but a very small amount of our overall DNA sequence—16,569 base pairs, to be exact— resides in our mitochondria and is called *mitochondrial DNA*. Our mitochondrial DNA mutates at a somewhat higher rate than our nuclear DNA, so the variation observed among people when their mitochondrial DNA sequences are compared is generally higher than when nuclear sequences are compared.

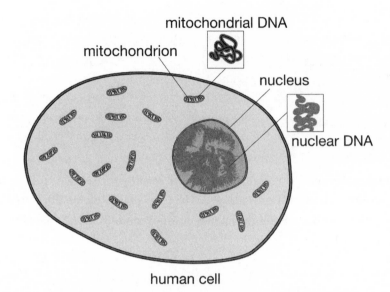

mitochondrial DNA

mitochondrion

nucleus

nuclear DNA

human cell

Figure 4.3. Nuclear DNA and mitochondrial DNA within a cell. Mitochondrial DNA is in the mitochondria, which reside outside of the nucleus in a cell.

Everyone has mitochondrial DNA, and everyone inherits it only from her or his mother. Even though both females and males have mitochondrial DNA, its line of inheritance is purely maternal—from mother to daughter—rendering the mitochondrial DNA in males a hereditary dead end. Thus, by comparing variations in mitochondrial DNA among people, scientists can trace pure maternal lineages.[5]

Another type of DNA that tends to vary more than the rest is the DNA in the Y chromosome in males. A gene on this chromosome serves as the master switch for sex determination—the presence of the gene nearly always results in development of a male, whereas its absence results in development of a female. Males have one Y chromosome and one X chromosome (XY), whereas females have two X chromosomes and no Y chromosome (XX). Males inherit their single Y chromosome only from their fathers. Thus, scientists can trace paternal lineages of inheritance by examining variations in Y-chromosome DNA.[6]

Mutations can happen in any person in any generation. They are rare, but once present in the DNA of an egg or sperm cell they can be transmitted to offspring. From that point forward, the mutation is called a *variant* in DNA, meaning that the DNA sequence varies among individuals because of the original mutation. The variant is replicated faithfully and can be passed on through an unlimited number of generations.

The following scenario is one that has happened independently many times: Thousands of years ago, a particular woman's mitochondrial DNA mutated, creating a new variant. She passed it on to at least one of her daughters, who passed it to her daughters, who passed it on to theirs, and so on to the present day. People whose maternal lineage traces back to this woman carry this variant in their mitochondrial DNA, whereas those who do not trace their maternal ancestry to this woman do not carry that variant. Generations later, a second mutation happened in a different place in the mitochondrial DNA in one of this woman's distant female descendants. That new mutation became a variant linked to the first variant but only in that second woman's descendants. Later, a third variant resulted from another mutation, so that there were three linked variants in the mitochondrial DNA in a particular maternal lineage, and later a fourth, then a fifth, and so on.

As human populations grow and people migrate to settle new areas, women pass on their mitochondrial variants from mother to daughter through many generations. The earliest variants should be present in a larger number of people spread over a larger geographic area than a variant resulting from a more recent mutation. And, because newer variants are superimposed on the background of older variants, scientists can determine approximately when a particular variant arose. By amassing and analyzing data from many of these accumulated variants, they can reconstruct maternal ancestries extending back many generations. If these variants are compared with geography—where the ancestors of the people who carry them lived—scientists can determine not only when but also *where* a particular variant arose, and in so doing they can reconstruct the migration and settlement patterns of ancient people.

The same is true for the Y chromosome. Mutations that happen in the Y-chromosomal DNA of men are variants transmitted to their sons, who transmit them to their sons, generation after generation. As later variants are superimposed on the background of earlier variants at particular times and in particular places, scientists can analyze these variants to trace when and where they occurred, and in this way they can reconstruct *paternal* ancestries.

Why is there so much focus on mitochondrial and Y-chromosome DNA when they constitute just a minor fraction of our DNA? There are two very good reasons. First, as mentioned earlier, mitochondrial and Y-chromosome DNA sequences tend to vary more than other types of DNA sequences, so there are proportionally more opportunities for scientists to trace different lineages with them. Second, we inherit nuclear DNA (except for the Y chromosome) from both parents, and variants inherited in the nuclear DNA from both parents can recombine—a variant inherited from your mother can be recombined with another variant inherited from your father. So the patterns of superimposition of variants, so readily evident in mitochondrial and Y-chromosome DNA, are often broken and reshuffled in our nuclear DNA. Although scientists are able to decipher evolutionary information from all types of DNA, the most informative and powerful stories of human migratory and settlement history come from mitochondrial and Y-chromosome DNA.

NUMEROUS RESEARCH PROJECTS ARE REVEALING ANCIENT HUMAN HISTORY THROUGH DNA ANALYSIS.

The idea that genetic variants can tell the story of ancient human migration and settlement is not new. Luigi Luca Cavalli-Sforza recognized the opportunity decades ago, well before automated DNA sequencing and analysis were available. In his book *The Great Human Diasporas*, he wrote, "Can the history of humankind be reconstructed on the basis of today's genetic situation? This was the question I posed to myself over forty years ago. I made a personal bet it could be done, because I believed the theory of evolution gives us the key.[7] Cavalli-Sforza and statistician A. W. F. Edwards analyzed a few nuclear genetic variants, such as blood types, identified among indigenous people throughout the world. Their research strongly pointed to Africa as the place where our species originated. Not surprisingly, this information independently confirmed the evidence from hominin fossils, as we saw in the previous chapter.

Later, as methods for analysis of DNA became available, scientists began examining mitochondrial and Y-chromosome DNA. This work has confirmed the general pattern of an African origin for humans followed by migration from Africa by people who became the founders of those who populated the rest of the world. The story is now greatly amplified with enormous amounts of detail. A significant part of current research on deciphering the ancient history of humanity, largely through DNA analysis, is centralized in the Genographic Project. This project is a major research effort sponsored by several organizations, including the National Geographic Society, IBM, the Waitt Family Foundation, and Applied Biosystems, and its website offers an excellent overview of how mitochondrial and Y-chromosome DNA reveal the ancient geographical history of humanity.[8]

Several research projects have amassed an enormous amount of information from DNA analysis and geography. The process is based on the fundamental concepts of haplotype and haplogroup, words you are likely to see when you read almost any source about ancient human migrations and DNA. A *haplotype* is a unique set of variants in DNA—essentially a specific genetic signature. Even a single variant in just one base pair of DNA distinguishes a particular haplotype from its most closely related haplotype. For example, if the region of DNA being examined is the entire mitochondrial DNA, everyone with a particular haplotype has exactly the same DNA sequence in their mitochondrial DNA. Any new mutation creates a new haplotype. As a consequence, there are many thousands of mitochondrial and Y-chromosome haplotypes.

A *haplogroup*, by contrast, is a grouping of related haplotypes that all carry the same *ancient* variants. Each haplogroup identifies a broad group of people who are genetically related to one another through their maternal or paternal lineages by distant ancestry. A new mutation that creates a new haplotype does not create a new haplogroup. The new haplotype remains as a member of its haplogroup because only ancient variants identify haplogroups.

To define a haplogroup, scientists must first distinguish ancient and modern variants derived from the original ancestral sequence. This original ancestral sequence can be statistically identified through comparison of sequences from many individuals and from closely related species. For example, molecular biologists have obtained complete mitochondrial DNA sequences from a large number of people whose ancient ancestry is spread across the world. They also have obtained complete mitochondrial DNA sequences from chimpanzees, gorillas, orangutans, and Neanderthals. If a particular base pair is present in a large subset of human mitochondrial haplotypes and is also present in the chimpanzee, gorilla, orangutan, and Neanderthal mitochondrial genomes, it must represent the original *ancestral* sequence at that site. Any variant in the human sequences present in just a subset of humans must have arisen by mutation at some point in human evolutionary history after divergence from the other groups and is a *derived* variant.

Each haplogroup is typically defined by a set of well-defined ancient derived variants spread throughout the DNA, rather than just by one or two variants. On occasion, a variant may mutate *back* to the original ancestral sequence, so the variant appears to be ancestral, when in fact it is new. Fortunately, scientists can distinguish new back-mutations from ancient mutations by their distributions—ancient mutations are widespread, whereas new mutations are present in smaller subsets of people. And the fact that a haplogroup is defined by a set of multiple ancient variants means that any new mutation, including a back-mutation, is superimposed on a well-defined genetic background. Back-mutations are very rare but can be identified and excluded when a haplotype containing one is assigned to a haplogroup.

DNA DIVERSITY OVERLAID WITH GEOGRAPHY IDENTIFIES WHERE A SPECIES ORIGINATED AND WHERE SPECIFIC HAPLOGROUPS ORIGINATED.

The fossil evidence, which we explored in the previous chapter, points to East Africa as the place where our species—*Homo sapiens*—originated, about two hundred thousand years ago. This conclusion is derived solely from the fossils, without any reference to DNA. We can also analyze DNA diversity, independent of the fossil record, to tell where and when a species originated. The fundamental premise is a simple one—*the greatest and most ancient DNA diversity is typically found in the place where a species originated.*

The reason for this premise is also simple. As a new species emerges and establishes itself, mutations accumulate in the species, and its genetic diversity grows over many generations. As subgroups of individuals migrate away from the original population to settle elsewhere, they carry just a subset of the genetic diversity with them, leaving the highest diversity in the region of origin. After these subgroups establish themselves in new areas, new subgroups migrate from them, carrying even less diversity. Thus, we should see the greatest genetic diversity in the region of origin and the least where distant populations settled after multiple waves of migration. Moreover, because scientists can distinguish ancient from recent variants, they can identify those people in whom the most ancient sequences reside. The place of their ancestry also typically points to the region of origin.

The same is true for haplogroups themselves. The highest diversity within a haplogroup should be in the place where it originated, and diversity then decreases as groups of people carrying the haplogroup migrate away from the center of origin. By collecting and analyzing mitochondrial DNA from large numbers of indigenous people throughout the world, then comparing haplogroup diversity with the places where these people live, scientists can often identify where a haplogroup originated and how people carried subsets of it elsewhere in the world. Combining information from all known haplogroups, they can ultimately reconstruct ancient human origins, migrations, and settlements.

Mitochondrial DNA variants show how the human mitochondrial genome has evolved from a single, most recent common ancestor.

Thanks to modern high-throughput DNA sequencing and a widespread scientific effort to collect DNA from indigenous people throughout the world, we now have an enormous amount of mitochondrial DNA sequence information from the world's human population. Scientists have taken this information and constructed—purely on the basis of DNA sequence with no reference to geography in the analysis—the evolutionary relationships of the various mitochondrial haplogroups.[9] These haplogroups are identified with uppercase letters, from A to Z, sometimes followed by numbers and lowercase letters to distinguish closely related haplogroups derived from the same original source.

The relationships that emerge from this analysis can be distilled into two key points: First, the most diverse and ancient haplogroups are all related to one another and are all labeled with the letter L, which represents a *superhaplogroup* because the haplogroups within it are so ancient and so diverse. In fact, there is more than twice as much diversity in the L haplogroups than in all other haplogroups combined.[10] Second, one of the L haplogroups, L3, was the source of two major haplogroups, M and N, from which all non-L haplogroups are derived.[11]

As we'll see throughout the rest of this chapter, when haplogroup relationships are overlaid on geography, a clear story of ancient human history emerges. The L haplogroups originated in sub-Saharan Africa (the part of Africa south of the Sahara Desert), with the most ancient haplogroups concentrated in East Africa. Thus, both the fossil evidence and DNA evidence independently point to East Africa as the place where modern humans originated. People carrying haplogroup L3 left Africa, and from L3 two haplogroups, M and N, arose in the Middle East. The multiple generations of people who populated the world outside of Africa carry haplogroups derived from M and N.

The accumulated evidence derived from mitochondrial diversity points to a single woman who was the first to carry the original L haplogroup. She lived in Africa about 192,000 years ago and is the common mitochondrial ancestor of all humans alive today.[12] This woman is popularly known as the *mitochondrial Eve*, and she has been sensationalized in countless articles, books, talk shows, and other venues. Often omitted is the fact that she is only our *mitochondrial* common ancestor, not the common ancestor for the rest of our DNA. In fact, because of the way nuclear DNA is inherited and recombined, there is a good chance that *only* our mitochondrial DNA came from her. Although she is our distant ancestor, her nuclear DNA has become so diluted that only the tiniest fraction, if any, of it remains in anyone alive today.

This discovery—that all of us trace our mitochondrial genetic origin to a single person—is not the extraordinary scientific breakthrough it might seem to be. The concept of tracing ancestry to a single individual is very old and well established in evolutionary theory. Such an individual is called the *most recent common ancestor*, often abbreviated in scientific discussions as MRCA. In the case of mitochondrial DNA, all of us trace our mitochondrial ancestry to one woman who acquired the final mutation that became part of the founding set of mitochondrial variants that distinguish human mitochondrial DNA from that of our closest evolutionary relatives. Her mother was also a mitochondrial Eve for all of humanity, as was her mother's mother, and so on, generation after generation. Ultimately, all humans, extinct hominins, chimpanzees, and bonobos collectively have a mitochondrial MRCA who lived more than five million years ago. There is another mitochondrial MRCA for all humans and great apes, another for all primates, and yet another for all mammals, and so on back to the origin of mitochondria more than a billion years ago.

Also, because the mitochondrial Eve is not the MRCA for the rest of our DNA, there must be multiple "Eves" and "Adams" in our ancestry. Ultimately, there are MRCAs for every base pair in our DNA, and those ancestors lived throughout a wide range of times. The fact that we share about 92 percent of our DNA sequence with Old World monkeys means that the vast majority of MRCAs for our DNA date to before the time when the ancestral human–great ape lineage and Old World monkey lineage diverged between twenty-four and twenty-nine million years ago.[13]

MITOCHONDRIAL HAPLOGROUPS REVEAL HOW ANCIENT PEOPLE POPULATED THE WORLD.

Let's explore what mitochondrial haplogroups tell us about the ancient history of humanity, beginning with the origin of modern humans and the L haplogroups in Africa. Haplogroup L0 is the most ancient of all mitochondrial haplogroups, and it originated in East Africa. Within the L0 haplogroup, the most ancient subgroup is designated L0d. This subgroup is found mostly in two populations—the Sandawe of Tanzania and the southern African Khoisan-speaking people.

This L0d subgroup is an excellent example of how information from DNA analysis, geography, anthropology, and language comes together to reveal ancient human history. Khoisan is one of four ancient language groups in

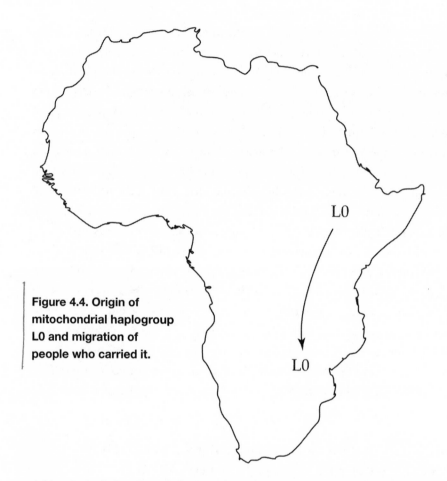

Figure 4.4. Origin of mitochondrial haplogroup L0 and migration of people who carried it.

Africa. It includes the "click" languages—people who speak them make distinct clicking sounds. The Sandawe and southern African Khoisan-speaking people speak related click languages, even though they are geographically separated. Combined evidence from DNA sequencing, geography, archaeology, and language suggests that the L0 haplogroup originated in East Africa about 150,000 years ago and that a group of people carrying it migrated south to found the southern population.[14]

The next most ancient African haplogroups are L1 and L2. Haplogroup L1 arose by mutation from L0 about 140,000 years ago in East Africa, and from that time forward, people carrying both L0 and L1 lived side by side. Eventually, some people bearing only L1 migrated west, and from them

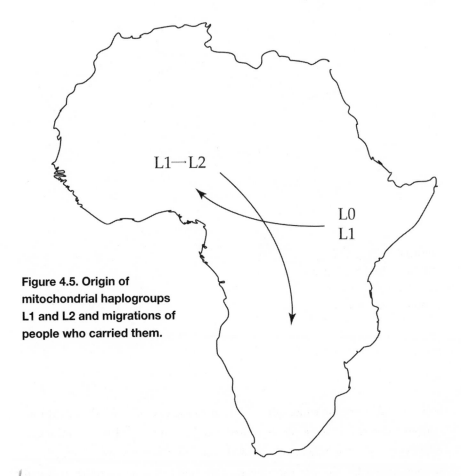

L1→L2

L0
L1

Figure 4.5. Origin of mitochondrial haplogroups L1 and L2 and migrations of people who carried them.

emerged another African haplogroup known as L2 in West Africa about ninety thousand years ago. This haplogroup is now highly diverse and distributed throughout Africa. It also is the most prevalent haplogroup among people whose ancestors were taken as slaves from Africa in more recent times. Its widespread dispersion in Africa, and its prevalence outside of Africa, make it the most common haplogroup among people whose maternal ancestry is African, such as African Americans (in both North America and South America) and African Europeans.

The next most ancient haplogroup is L3, which is a fascinating haplogroup because all people whose ancient ancestry is outside of Africa trace their mitochondrial origins to it. Haplogroup L3 arose from L2 in Africa around 72,000

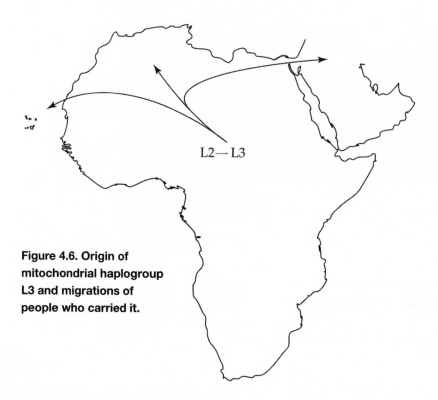

**Figure 4.6. Origin of
mitochondrial haplogroup
L3 and migrations of
people who carried it.**

years ago. About that time, rainfall began to increase in the Sahara, turning
it from a desert into a wetter, vegetated area.[15] Animals and people migrated
northward into the Sahara region, and some of these northward-migrating
people carried L3. Groups carrying L3 continued to migrate northward and
westward, as far as the Cape Verde Islands off the west coast of Africa. At least
one group of people migrated eastward, leaving the African continent and settling in the Arabian Peninsula, where some of the group's descendants still live.

As we saw in the previous chapter, these people were not the first hominins to leave Africa—Neanderthals, the descendants of a much earlier migration, were already widespread in Europe and the Middle East when humans
began to migrate out of Africa. Haplogroup L3–bearing people also were not
the first of our species to leave Africa. Remains of a human population dating
back to about 100,000 years ago were found in the Qafzeh and Skhul Caves in
Israel. These people, however, disappear from the fossil record at about 70,000
years ago, and they apparently left no descendants.

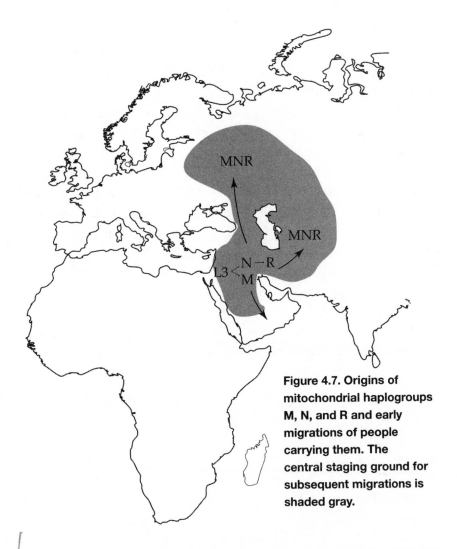

Figure 4.7. Origins of mitochondrial haplogroups M, N, and R and early migrations of people carrying them. The central staging ground for subsequent migrations is shaded gray.

Two haplogroups, M and N, arose from L3 between 70,000 and 50,000 years ago, and shortly thereafter a third haplogroup, called R, arose from N. Like L, these three haplogroups—M, N, and R—are *macrohaplogroups* because they became the ancestral haplogroups of all people who lived outside of Africa for tens of thousands of years, as generations of people populated the rest of the world.

People bearing macrohaplogroups M, N, and R expanded into the Middle East and eventually into a large region in Central Eurasia (depicted in figure

4.7, page 113), which encompasses much of the Middle East and areas north, including the region surrounding the Caspian Sea, the Caucasus Mountains, and what are now Western Russia and the Baltic states. Many of the haplogroups found throughout the world arose from M, N, and R in this region, which became a central staging ground for migrations east into Asia and west into Europe, as well as some migrations south, returning into the Arabian Peninsula and back into Africa.

People carrying haplogroup M and their descendants migrated mostly eastward out of this region and settled widely throughout central and southern Asia. In one southern migration route, people moved through what is now India into Southeast Asia and ultimately into Australia, where members of aboriginal populations today carry subgroups of M. Numerous subgroups of haplogroup M are widespread among indigenous people, ranging from the Arabian Peninsula in the west and central Asia in the north to Australia in the southeast.

As mutations accumulated in haplogroup M, several important haplogroups arose from it east of the Caspian Sea. Among these were haplogroups

Figure 4.8. Migrations of people carrying haplogroup M.

Figure 4.9. Origins of mitochondrial haplogroups C, D, and Z and migrations of people carrying them.

C, Z, and D, which are closely related. They arose about 26,000 to 24,000 years ago when a single haplogroup arose from M, then later diverged into C and Z. Haplogroup D is more distantly related to C and Z, and it arose from M in the same general area but much earlier, about 48,000 years ago.

As shown in figure 4.9, above, people carrying C, Z, and D and their descendants migrated north and east from central Asia, settling in and crossing ancient Siberia. Currently, about 20 percent of people who are native to Siberia carry haplogroup C, and another 20 percent carry D, whereas haplogroup Z accounts for only about 3 percent of people in this region.[16]

Groups bearing C and D migrated from Siberia to the south, west, and east. In the west, haplogroups C and D dwindle where the Ural Mountains were a barrier to migration. Migrations to the south in Asia also dwindled the farther south they proceeded. To the east, however, the migrations of C and D would continue for an enormous distance.

Between about 22,000 and 15,000 years ago, during a dry climatic period and ice age, sea levels were much lower than they are now because much of the water currently in the oceans was trapped in polar ice caps. What is now the Bering Strait between northeastern Siberia and western Alaska was then a broad land bridge named Beringia, which connected Asia with North America, where the Pacific Ocean kept the southern shoreline warm enough to be free of ice. People could migrate along it, feeding themselves by hunting and fishing, much as those who now live in this region do.

Around the latter end of this ice age, as the climate began to warm, about 15,000 years ago, people with haplogroups C and D and their descendants crossed through Beringia into what is now Alaska and migrated southward throughout North America, eventually continuing into Central America and South America. The greatest diversity within haplogroups C and D is in central Asia and Siberia, and the least is in South America, indicating that the genetic diversity of these haplogroups decreased with each wave of migration from north to south in the Americas. Haplogroups C and D are now two of the most widely dispersed mitochondrial haplogroups in the world, ranging from central Asia, through Siberia and North America into South America.

Few archaeological relics remain from these ancient migrations across Beringia and along the western coast of North America. The shoreline these people apparently used for migration is now under water, inundated by the Pacific Ocean when polar ice melted at the end of the ice age. From that point forward, Beringia was no longer a land bridge. Instead it became the Bering Strait, a barrier to further migrations, which had become more difficult because they required navigation on seaworthy boats across waters that were then (and are now) often cold, turbulent, and dangerous.

People carrying haplogroup Z also migrated into Siberia along with those carrying C and D. Their dispersion, however, was much more limited, and their populations were smaller. Some migrated westward, but their migration dwindled at the Ural Mountains. Others migrated east through Siberia (where most people with haplotype Z are now found) and south into China. Haplogroup Z is purely an Asian haplogroup. If it ever made its way into the ancient Americas, it has long since disappeared. In Asia and Siberia, it is not nearly as widespread as its close cousin, haplogroup C, or its more distant cousin, haplogroup D.

Three other haplogroups, G, Q, and E, arose from M. Haplogroup G is present in a small percentage of people in eastern Siberia and east Asia and did not spread widely beyond this region. Haplogroup Q arose about 50,000 years ago,[17] and E, about 35,000 years ago[18]—both from subgroups of M in Island Southeast Asia, a region that now encompasses Indonesia, Malaysia, the Philippines, and Papua New Guinea.

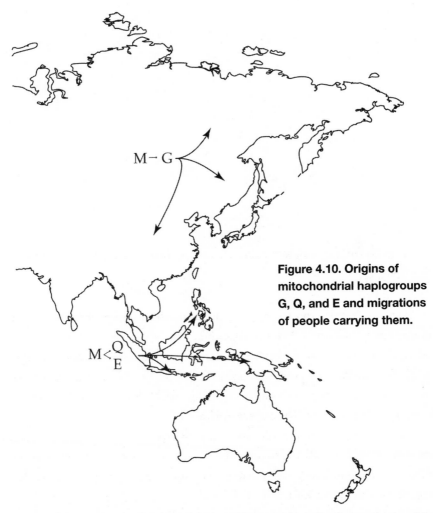

Figure 4.10. Origins of mitochondrial haplogroups G, Q, and E and migrations of people carrying them.

Ancient people carrying these haplogroups remained mostly in this region, their expansion coinciding with the last ice age. Although this area was in a tropical region and free of ice year-round, sea levels were much lower than they are now. The islands of Sumatra, Borneo, and Java were joined to mainland Asia in a large landmass extension of the Asian continent known as Sundaland. Eventually, with the end of the last ice age between 15,000 and 7,000 years ago, sea levels rose, and much of the lowland occupied by people carrying these haplogroups was inundated with seawater, resulting in the formation of the islands now in this region.

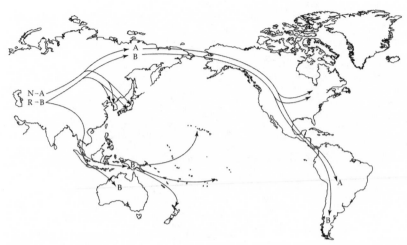

Figure 4.11. Origins of mitochondrial haplogroups A and B and migrations of people carrying them.

Let's now return to the staging ground in central Eurasia and follow the fate of macrohaplogroup N and its daughter, macrohaplogroup R. People carrying N did not spread far before it began to mutate into new haplogroups. Their descendants, who carried a large number of haplogroups derived from N, populated much of the world, including the Middle East, Asia, Europe, and the Americas, as well as parts of Africa through back-migration into that continent. The haplogroups that arose directly from N in this region include A, I, W, Y, S, X, and R. Haplogroup R was especially important because it, in turn, was the source of some of the most widespread haplogroups in the world, among them U, K, HV, H, V, B, F, J, and T.

Haplogroup A is a large, diverse, and widespread haplogroup. It arose from N in the central Eurasia staging ground east of the Caspian Sea about 29,000 years ago. People bearing this haplogroup moved eastward, and their descendants are now widespread in China, Japan, Korea, and Siberia. Around 15,000 years ago, a group of people in northeastern Siberia who carried haplogroup A migrated through Beringia into North America. Distinctly American subgroups of haplogroup A evolved in the Americas and are now found among indigenous people from Alaska to South America.

Haplogroup B is one of the oldest, most diverse, and most widespread of all mitochondrial haplogroups outside of Africa. It was one of the first haplogroups to emerge from R, about 51,000 years ago, somewhere east of the Caspian Sea, and it was one of the first to enter Asia. It spread into Southeast Asia, Island Southeast Asia, and the Pacific Islands. It also was one of five haplogroups (A, B, C, D, and X) present in people who crossed Beringia about 15,000 years ago into North America. From there, people carrying it migrated throughout both North America and South America, and it is now the most prevalent haplogroup among native people in the southernmost regions of South America. Various subgroups of haplogroup B are now dispersed among indigenous people throughout China, Japan, Southeast Asia, the Pacific Islands, North America, and South America (figure 4.11, page 118).

Like B, haplogroup F is old, having arisen from R about 43,000 years ago east of the Caspian Sea. People carrying this haplogroup spread eastward throughout Asia and into Siberia, and southward through Southeast Asia into Island Southeast Asia in what is now Indonesia, the Philippines, Borneo, and

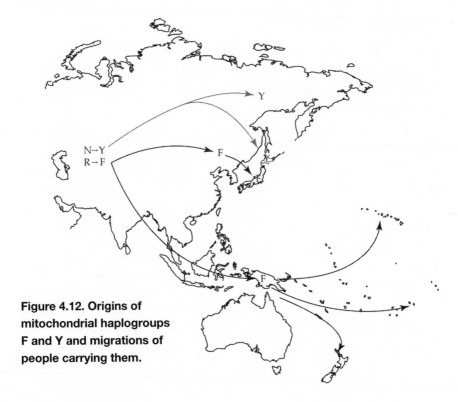

Figure 4.12. Origins of mitochondrial haplogroups F and Y and migrations of people carrying them.

Papua New Guinea. People bearing this haplogroup also migrated eastward into the Pacific Islands, where haplogroup F is present in a small proportion of people. In contrast to these widespread groups, haplogroup Y has very limited distribution, being found among certain groups of indigenous people in what is now eastern Russia, Japan, and Sakhalin Island. It arose more recently, around 22,000 years ago, in east Asia.

Let's now turn our attention to those haplogroups found predominantly west of the central Eurasian staging ground. One of the earliest haplogroups

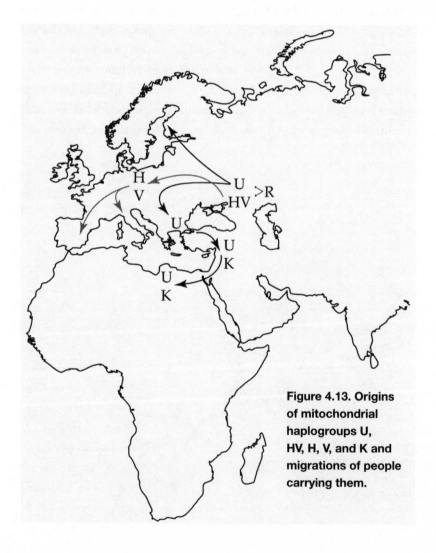

Figure 4.13. Origins of mitochondrial haplogroups U, HV, H, V, and K and migrations of people carrying them.

to originate from R in this region was U, which arose about 55,000 years ago. People bearing U migrated north into what is now Finland and Norway, west into eastern Europe, and a considerable distance south into North Africa. Because U is so old, several haplogroups were derived from it, designated as U1 through U8 and K. For example, U5 is a subgroup that arose in people migrating northward and is now most prevalent in Finland. Haplogroup K arose from U8 during a southern migration and is currently found in Syria, Lebanon, Jordan, Israel, North Africa, and parts of Asia and southern Europe.

Another haplogroup to arise from R was HV, probably in the Caucasus Mountains about 30,000 years ago. People bearing this haplogroup migrated northward into the Baltic region and eastern Europe, where two related haplogroups, H and V, diverged from it. People with H and V migrated westward into northern Europe. Then, about 20,000 years ago, during an ice age, descendants of these people moved southward into the warmer climates of southern Europe, including the Iberian Peninsula, southern France, Italy, and the Balkans. When the climate warmed and the northern ice retreated, around 15,000 years ago, people from the south who carried H and V moved north, back to the lands where their ancestors had previously lived. Others bearing H and V remained in the south. As a consequence, H and V are the most prevalent and widespread haplogroups throughout Europe, found in about 75 percent of all people whose maternal ancestry is European (figure 4.13, page 120).[19]

Also in the Caucasus Mountains and the region north of there, two haplogroups, I and W, arose directly from N, probably about 30,000 years ago. People carrying I migrated northward into what are now the Baltic states and Russia and westward into Europe. Haplogroup I is rare, usually present in less than 4 percent of people in the regions where it is found. Haplogroup W also was carried westward into Europe, but, unlike haplogroup I, it also made its way eastward into a region that now includes Iran, Pakistan, and Afghanistan. It is rare wherever it is found.

A younger set of haplogroups, J and T, are closely related to each other, both having arisen from R near modern-day Lebanon about 10,000 years ago. They spread widely through migrations from their place of origin, northwest into Europe, north into the Baltic region, southwest into the Arabian Peninsula, and east into what is now Pakistan. People who carried J and T

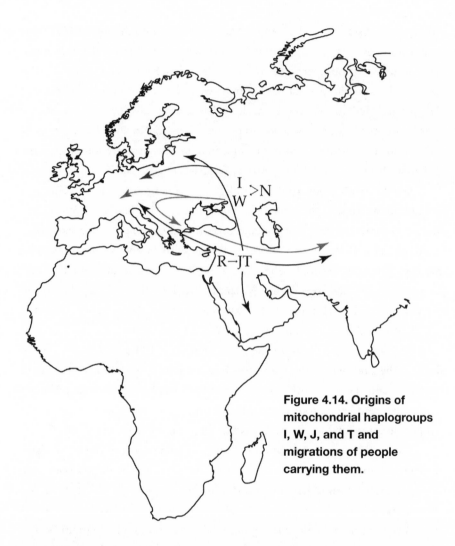

Figure 4.14. Origins of mitochondrial haplogroups I, W, J, and T and migrations of people carrying them.

were probably descendants of the primitive agriculturalists who domesticated wheat about 10,000 years ago in the Fertile Crescent region of what is now Iraq, Lebanon, Turkey, Syria, and Israel (figure 4.14, above).

The most enigmatic mitochondrial haplogroup is X. It is quite rare, nearly always a minority haplogroup, and highly dispersed, found in a few populations scattered across Europe, the Middle East, North Africa, East Africa, Siberia, and the Americas. It is most prevalent in populations of the Druze in the Middle East and the Ojibwa in North America. In these two populations,

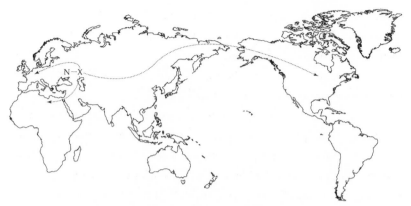

Figure 4.15. Most probable origins of mitochondrial haplogroup X and likely migrations of people carrying it.

about one of every four people carries it; elsewhere the proportion of people carrying it tends to be small. It arose from N about 32,000 years ago, probably in the same central Eurasian staging ground as most other haplogroups derived from N. Figure 4.15 shows the most likely migration routes: west into Europe, south through the Middle East and into Africa, and east across Asia, through Siberia, and into North America.

The four other American haplogroups (A, B, C, and D) are widespread throughout North America and South America, whereas haplogroup X is restricted to a few North American native populations—the Ojibwa, the Sioux, the Nuu-chah-nulth, and the Navajo. Haplogroups A, B, C, and D are also prevalent in Siberia, and the patterns of variants in them offer strong evidence that all four entered the Americas from Siberia through Beringia. For some time, haplogroup X was thought to be absent from Siberia, having entered North America by an ancient migration from northern Europe. This conjecture, however, appears to be increasingly unlikely. Haplogroup X has since been found in the Altaians, a native Siberian population, suggesting a possible Siberian origin for the American subgroups of X.[20] Nonetheless, no subgroup of X anywhere in the world, including the Altaian subgroup, is clearly ancestral to the distinctly North American subgroups.

The degree of genetic diversity in the North American subgroups of X is essentially the same as that in the American subgroups of A, B, C, and D, strong evidence that all five haplogroups entered the Americas at about the same time through Beringia about 15,000 years ago, although not necessarily through a single migration event.[21]

Difficulties in tracing the lineages of haplogroup X are probably a consequence of *genetic drift*, a random natural phenomenon that often causes extinction of rare genetic types. Throughout nearly all of human history, a high proportion of people died before reproducing. If a particular haplogroup was rare in a population of people, there was a good chance of it being lost from that population purely by chance. As an analogy, imagine a jar with 100 marbles, 95 red and 5 green. If you reach into the jar without looking and randomly choose five marbles, discarding the rest, there's a good chance that you'll end up with only red marbles—all the rare green ones discarded. For this same reason, the few women who carry a rare sub-haplogroup in a particular generation might not leave any offspring who survive and reproduce, so the sub-haplogroup is lost forever in the descendants of those people. This probably happened in Siberia to the X sub-haplogroup that was ancestral to the North American X sub-haplogroups. Or, there remains the possibility that the ancestral type is still present, but because of its rarity it has not yet been found. Today, a few X sub-haplogroups are scattered throughout various places in the world as remnants of rare ancestral lineages possibly lost through genetic drift.

Y-CHROMOSOME HAPLOGROUPS TELL MUCH ABOUT BOTH ANCIENT AND RECENT HUMAN HISTORY.

Because Y chromosomes are transmitted only from father to son, they provide a way for scientists to trace pure *paternal* lineages. As was the case with mitochondrial DNA, all human Y chromosomes trace their origins to a most recent common ancestor, sometimes called the Y-chromosome "Adam." The mitochondrial Eve and Y-chromosome Adam both lived in east Africa but the times when they lived were separated by tens of thousands of years. As we saw earlier, the mitochondrial Eve lived about 192,000 years ago. Although it is more difficult to pin down the time when the Y-chromosome Adam lived,[22] it was probably as recent as about 70,000 years ago.[23] Because of the dilution of any one ancestor's nuclear DNA with every generation, it is highly unlikely that anyone on the planet today retains any DNA inherited from him, except for the Y-chromosome DNA, which all human males inherited from him through many generations.

His original Y chromosome has since diverged into twenty major haplogroups designated by the letters A through T.[24] The overall migration patterns deduced from these haplogroups and their geographic distributions are similar to those identified from mitochondrial DNA, although they are neither as detailed nor as well refined. As expected, the most ancient and diverse Y-chromosome haplogroups are in Africa. Thus, Y-chromosome haplogroups offer yet further evidence of an African origin for humans.

The same staging ground in central Asia for mitochondrial haplogroups was the source of Y-chromosome haplogroups that were eventually dispersed throughout the world outside of Africa. We'll not revisit in detail the similarity of mitochondrial and Y-chromosome migration patterns here. Instead, we'll focus on some of the unique aspects of human history revealed by Y-chromosome DNA when overlaid on geography.

To begin, let's return to the Roman myth of Romulus and Remus. At the beginning of this chapter, we left the story shortly after Romulus had killed his brother Remus and founded the city of Rome. According to the myth, Romulus recruited a group of men who joined him in building Rome. In need of wives, they sought women among the Sabines, who lived in a region of Italy northeast of Rome. The Roman men forcibly abducted the Sabine women, taking them to be their wives. For centuries, the abduction of the Sabines (sometimes called the rape of the Sabines) has been a favorite theme in European art.[25]

Turning to another myth, most people are familiar with the classic Trojan Horse story from Virgil's *Aeneid*. As a strategy to end its years of war with Troy, the Greek army resorted to a ruse. Its forces began to sail away from Troy in an apparent retreat, leaving a large wooden horse at the gates of Troy as a gift. The Trojans pulled the horse into the city, but, unbeknownst to them, hidden within the horse was a small contingent of Greek soldiers. In the dead of night, the hidden soldiers crept from within the horse and opened the city gates for the returning Greek army to invade. The Greeks slaughtered the Trojan men and took women and children as spoils of the war.

Although these stories are purely mythical, they portray an ancient practice in which women are kept alive during an invasion, and some of them bear offspring fathered by their captors, a pattern that unconscionably and tragically still happens today (the Bosnian and Rwandan wars are recent examples). The genetic consequence is preservation of the mitochondrial DNA in the conquered people but introduction and spread of the Y-chromosome DNA by the invading men. The technical term for the presence of two genetically distinct types in a population is *admixture*, and it results from matings between females and males from different genetic backgrounds. There is ample evidence from mitochondrial and Y-chromosome studies that such admixture events have repeatedly happened throughout human history and have shaped the genetic landscape of modern humanity.

Let's look at four examples, the first of which comes from Africa. The most ancient Y-chromosome haplogroups are A and B. Not surprisingly, A, which is the most ancient, is most prevalent in people belonging to indigenous Khoisan "click"-language populations in eastern and southern Africa, the same groups that carry the most ancient mitochondrial haplogroups. The next most ancient is haplogroup B, which arose from A. It, too, has a limited distribution, mostly in hunter-gatherer tribes of Africa. Today it is most prevalent among African pygmy populations.[26]

As a general rule, the most ancient haplogroups are also the most diverse, and at one time this was probably true for Y-chromosome haplogroups A and B in Africa. However, haplogroup E is now the most widely distributed and most diverse Y haplogroup in Africa. In fact, it is substantially more diverse than any other Y-chromosome haplogroup in the world. According to the latest data available, about one-sixth of the world's known Y-chromosome diversity resides in haplogroup E, most of it in sub-Saharan Africa.[27]

The reason why E is more diverse than the more ancient A and B haplogroups is the Bantu expansion. The Bantus were a group of people who lived in the grasslands of central Africa and developed an agricultural society in the midst of hunter-gatherer societies surrounding them. About four thousand years ago, they began an expansion that would last for more than a thousand years, displacing the hunter-gatherer societies present throughout much of sub-Saharan Africa. Bantu males carried the E haplogroup, and it now predominates throughout most of sub-Saharan Africa where the Bantu expansion took place.

This change in diversity is also reflected in African languages. The Khoisan click languages are confined mostly to people who still live in hunter-gatherer societies, and they tend to carry Y-chromosome haplogroups A and B. Bantu languages, by contrast, are diverse and widespread throughout sub-Saharan Africa. There are now more than five hundred Bantu languages and dialects, and most of the men who speak them carry haplogroup E.

The second example is the conquest of Genghis Khan, whose empire at the time of his death in 1227 CE stretched from the Pacific Ocean in the east to the Caspian Sea in the west throughout what are now southern Russia, Mongolia, northern China, Kazakhstan, Uzbekistan, Turkmenistan, and northern Afghanistan. Genghis Khan, along with his male descendants and male relatives, left their Y-chromosome throughout this region, which belongs to sub-haplogroup C3. Currently, more than sixteen million men (about 0.5 percent of all men in the world) carry this Y-chromosome sub-haplogroup, most of them still residing in the region of Genghis Khan's ancient empire.[28]

We now move to the Andes Mountains of Peru, where in 1532 Francisco Pizarro conquered the Inca Empire with a contingent of fewer than two hundred Spanish soldiers, opening the region to Spanish colonization. A 2001 survey of native Andean people in Peru showed that 94.2 percent of mitochondrial haplogroups were native (A, B, C, and D), whereas only 44 percent of Y chromosomes were native. The majority of Y chromosomes were from European haplogroups, originally from the invading soldiers.[29]

Our final example is from the United States. In 2006, scientists at the University of Arizona published a survey of men in the United States from four ethnic groups: African American, European American, Hispanic American, and Native American. The Native American group showed a clear trend. Eighty-nine percent of the Y chromosomes in Native American men from the northeastern United States were European—not coincidentally from the region where Europeans first colonized North America. The proportion of European Y chromosomes depended on how far west the Native American ancestries were—the smallest proportion was in Apache men from Arizona (7 percent).

Among African American men, 26.4 percent had European Y chromosomes, some a consequence of admixture more than a century ago during the period of slavery. Hispanic American men had mostly European Y chromosomes at 77.8 percent, but with significant Native American (13.7 percent) and African (6.6 percent) Y chromosomes. In sharp contrast, admixture was minimal among European Americans, with 96.6 percent European Y chromosomes.[30]

Studies from the Y chromosome have also helped clarify questions from mitochondrial studies. For example, Native American Y chromosomes show a distinct migration pattern by ancient people who crossed Beringia to become the founding population of Native Americans. And they further support a Siberian origin for mitochondrial haplogroup X in the Americas. Among Native Americans throughout the Americas, a single subgroup of Y-chromosome haplogroup Q predominates—subgroup Q3. According to a large-scale survey published in 2004 by scientists at the University of Arizona, 76.4 percent of Y chromosomes in Native Americans (from throughout the Americas) are from subgroup Q3. The next most common are American subgroups of haplogroup C at 5.8 percent, with nearly all others Y-chromosome haplogroups (17.3 percent) from European admixture.[31] Tracing the specific variants in the American sub-haplogroups of Q and C back to the Old World, these scientists determined that the probable origin of both haplogroups lies in the Altai Mountains of southern Siberia, where the modern-day borders of China, Russia, Mongolia, and Kazakhstan meet. This is also a region where all mitochondrial haplogroups in the Americas are likewise found in Asia. According to the authors of this study, "As far as we are aware, only the Altai region possesses all of the major Native American Y chromosome and mtDNA [mitochondrial DNA] founding haplogroups, thereby making it the best available candidate for the ancestral source region for the Native American population system."[32]

YOU CAN DISCOVER SOME OF YOUR OWN ANCESTRY THROUGH DNA ANALYSIS.

Everyone carries mitochondrial DNA and is a member of a mitochondrial haplogroup. You can find out your own mitochondrial haplogroup by having your DNA tested, then using this information to discover your maternal ancestral lineage. Several private and nonprofit laboratories can identify your mitochondrial haplogroup from your DNA for a fee. You simply visit the company's website and request a kit online, which arrives with instructions for swabbing the inside of your cheek to collect enough cells for DNA testing. You then return the swab to the laboratory, and after a few weeks, the laboratory will send you the results. To find a laboratory, search online for the words *mitochondrial*, *DNA*, and *testing*. Several laboratories offer the opportunity for you to voluntarily add your results to databases for research on mitochondrial haplogroup evolution.

If you are male, you can also have your DNA tested to determine your Y-chromosome haplogroup. The same companies that conduct mitochondrial DNA testing typically offer Y-chromosome tests as well, usually for an additional fee.

DNA testing has now become so routine that several companies offer genome-wide DNA tests that detect tens to hundreds of thousands of DNA variants throughout a person's DNA. The results of such tests can identify details about your ancestry that go beyond the mitochondrial DNA and Y chromosome, exploring multiple lineages. These tests can also reveal whether or not you are a carrier of some variants that cause genetic disorders, and they can give some indication regarding the effectiveness of certain medications given your particular genetic background. As researchers learn more about the genetic basis of medicine, DNA testing will increasingly become a valuable tool in medical diagnosis and treatment.

Our discussion of mitochondrial and Y-chromosome DNA in this chapter is just the tip of the iceberg of what our DNA can tell us about our evolution. In the next chapter, we'll look more deeply into the DNA of our chromosomes, showing how our chromosomes evolved and are continuing to evolve, and revealing what they tell us about our evolutionary relationships with other species.

CHAPTER 5
EVIDENCE FROM OUR GENOME
EONS OF SHUFFLING AND REARRANGING

Mythology is replete with fantastical hybrid creatures. Some are human-animal hybrids, such as mermaids, minotaurs, centaurs, fairies, sphinxes, fauns, and cupids. Others are animal hybrids, among them the hippocampus (sea-horse), griffin (eagle-lion), allocamelus (donkey-camel), chimera (fire-breathing lion-goat-snake), and Pegasus (winged horse). They are the stuff of pure fantasy—the whimsical creations of human imagination—their implausibility enhancing their mystique. They never existed beyond the confines of human fantasy, although many people believed in them in ancient times, and some still do today.

Real hybrids do exist, although they are not nearly as incredible as those of mythology. Within a species, hybrids are the first-generation offspring of parents who are genetically distant from each other, and within-species hybrids are almost always fertile. In some cases, a genetic phenomenon known as *hybrid vigor* confers exceptional growth to hybrid offspring. Hybrid corn, which is grown in nearly all commercial cornfields, is fast-growing, tall, and vigorous compared to nonhybrid corn. The same is true of hybrid tomatoes, squash, onions, cucumbers, cabbage, and many other foods.

Some closely related species may hybridize and produce what is known as *interspecific-hybrid offspring*, such as mules, which are horse-donkey hybrids. Interspecific hybrids are possible only between very closely related species, and often the hybrid offspring are sterile, though not always. Commercial sugarcane, which is the source of most of the world's sugar, is derived from interspecific hybrids between domesticated sugarcane and a related wild grass species, and these hybrids vary widely in their fertility. As briefly highlighted in chapter 1, lions and tigers are separate species, but they blur the line. Lion-tiger hybrids are called ligers, and several of them have lived to old age. In the wild, the ranges of lions and tigers do not overlap, so they do not naturally mate with each other—all known ligers are from matings in captivity. Male ligers are entirely infertile, incapable of producing offspring. Female ligers, on the other hand, retain some fertility; they are capable of producing offspring when they mate with a male lion. The same is true to a lesser extent for mules. Although most mules, male and female, are completely infertile, there are a few exceptionally rare documented cases of female mules giving birth to off-spring after mating with a male horse or donkey.

Interspecific hybrid ligers and mules illustrate the difficulty scientists have identifying what constitutes a species. We humans are clearly a distinct species with no blurred lines. The reason? Our closest living evolutionary relatives today are the common chimpanzee and the bonobo, and they are separated from us by at least five million years, which is more than ample time for different species to diverge from common ancestry. If Neanderthals were still here, the line might not be so distinct.

What causes this gradual breakdown of fertility as groups diverge from common ancestry to become separate species? A large number of biological factors are at play, including mating behavior and attraction, genes that evolve to confer reproductive incompatibilities, and changes in chromosomes. This latter phenomenon—changes in chromosomes—is one of the most important factors, and, under the microscope, it offers some of the most visible and com-pelling evidence of human evolution.

HUMAN AND CHIMPANZEE CHROMOSOMES DIFFER BY TEN REARRANGEMENTS.

As we've seen, our closest living evolutionary relatives are the common chimpanzee and the bonobo. In the cells of all organisms, DNA molecules are organized into chromosomes, each of which is one DNA molecule wrapped around a scaffold of proteins that stabilize it. When we compare human and chimpanzee chromosomes, ten major differences stand out—all of them large rearrangements of DNA. To highlight the evolutionary similarity between humans and chimpanzees, news reports often state that our DNA is more than 98 percent identical with chimpanzee DNA. At face value, this statement is correct; throughout most of our genome, the DNA sequence is about 98 percent identical to the chimpanzee genome. Within most genes—the most active parts of our DNA that govern the traits we carry—the identity is even higher, more than 99 percent.

But such a statement fails to tell the whole story. We must take into account not only the similarity of DNA sequences but also *rearrangements* of these sequences when related species are compared. In this chapter, we'll focus on those rearrangements and why they are so important for evolution.

Recall from the previous chapter that DNA is composed of four bases—T, C, A, and G. In the double-stranded DNA molecule, each A is paired with a T, and each G, with a C, so the molecule consists of A-T and G-C pairs. In the early 1970s, scientists stained human chromosomes with a chemical called giemsa and examined them under powerful microscopes. Giemsa darkly stains regions rich in A-T pairs but only lightly stains regions rich in G-C pairs. The result is an alternating light-and-dark banding pattern on each chromosome that under a microscope looks like a supermarket barcode. Every giemsa-stained chromosome has its own distinct barcode pattern—called a G-banding pattern in scientific terms—which allows scientists to identify each chromosome and distinguish it from all the others.[1]

Human chromosomes are designated by number as chromosome 1 through chromosome 22, plus the X and Y chromosomes, on the basis of their G-banding patterns.[2] And, except for the X and Y chromosomes in males, we have two copies of each chromosome in our cells, one copy inherited from each parent. The result is a total of forty-six chromosomes in our cells, twenty-three inherited from each parent. Thus, each of our cells (except for sperm and egg cells and a few other cell types) has essentially two genomes, a maternal genome inherited from the mother with twenty-three chromosomes, and the paternal genome inherited from the father, also with twenty-three chromosomes.

When scientists compared G-banded chromosomes from humans with

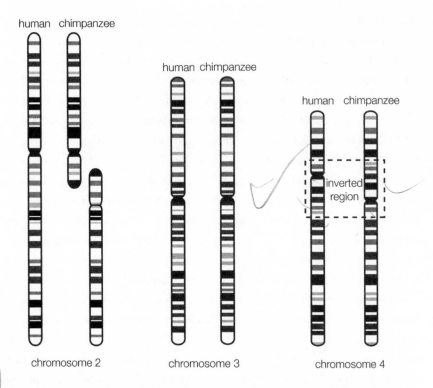

Figure 5.1. Comparison of G-banded human chromosomes 2, 3, and 4 with their chimpanzee counterparts.

those from chimpanzees in 1980, the similarities were immediately obvious—all human and chimpanzee chromosomes could be readily aligned according to their matching G-banding patterns.[3] For example, figure 5.1 compares G-banded human chromosomes 2, 3, and 4 with their chimpanzee counterparts. Notice that chromosome 3 has an identical G-banding pattern in both species, whereas chromosomes 2 and 4 are identical throughout large portions but with some significant rearrangements.

The DNA of the human genome was sequenced and assembled in 2001,[4] and the chimpanzee genome, in 2004,[5] allowing for a side-by-side comparison of DNA sequences between the two species down to the individual base pairs. The alignment of DNA sequences confirmed in much higher detail the original chromosome alignments from the 1980s. Such strong similarity between human and chimpanzee chromosomes, revealed both by chromosome banding patterns and DNA sequencing, constitutes some of the most powerful evidence of our common evolutionary ancestry with chimpanzees.

The differences in chromosomes 2 and 4 (as seen in figure 5.1, page 134) are two of ten major rearrangements that stand out when human and chimpanzee chromosomes are aligned. The most striking difference among all the chromosome comparisons is human chromosome 2, which matches *two* chimpanzee chromosomes because, long ago in human ancestry, two chromosomes fused into one. The remaining nine differences are *inversions*—large segments of DNA consisting of hundreds of thousands to millions of base pairs that are inverted in one chromosome when compared to its matching partner. Chromosome 4, also depicted in figure 5.1, is an excellent example of an inversion that distinguishes a human chromosome from its chimpanzee counterpart.

The sequencing of DNA has revealed, often in high detail, how each of these differences arose during evolutionary history. Let's look first at how the fusion in human chromosome 2 originated, then we'll look at two of the nine inversions as examples of how evolution has rearranged human and chimpanzee chromosomes.

HUMAN CHROMOSOME 2 AROSE FROM A FUSION OF TWO ANCESTRAL CHROMOSOMES.

In 2009, Elizabeth Blackburn, Carol Greider, and Jack Szostak were honored with the Nobel Prize in Physiology or Medicine for an astounding discovery they made in 1985. It has long been known that the DNA molecules in human chromosomes, and in the chromosomes of all other species except bacteria, are linear. Everything biologists knew about DNA and how it replicates seemed to indicate that linear chromosomes should erode on the ends every time they replicate. In fact, this is exactly what happens in many of our cells—the chromosomes *do* erode as these cells divide, and this limits the number of divisions many of our cells can make. This process of chromosome erosion is one of several factors that inevitably cause us to age. Throughout most of our bodies, our chromosomes and our cells are not immortal. They will age and ultimately die.

In some cells, however, chromosomes must persist through cell divisions, and in these cells the chromosomes must be protected from erosion. This is especially true for cells in the germline—those cells in our reproductive organs that produce egg and sperm cells. The chromosomes in these cells must be protected from erosion to be passed on from one generation to the next. Otherwise our chromosomes would have eroded into oblivion eons ago. Instead, something in the germline must either prevent the erosion or restore the eroded ends to maintain the integrity of our DNA. Greider and Blackburn, in a landmark paper published in 1985, reported what this something is.[6]

On the ends of every chromosome are specialized DNA sequences called *telomeres*. In humans, telomeres contain the same six-base-pair sequence

```
TTAGGG
AATCCC
```

repeated hundreds of times in tandem.

```
...TTAGGGTTAGGGTTAGGGTTAGGGTTAGGGTTAGGGTTAGGG... etc.
...AATCCCAATCCCAATCCCAATCCCAATCCCAATCCCAATCCC...
```

A protein called *telomerase* adds multiple copies of this six-base-pair sequence to the ends of chromosomes in the germline. When the chromosome replicates, the ends where the telomeric repeats reside begin to erode, but telomerase simply adds more copies, restoring the repeats and repairing the erosion. The restored ends protect the chromosome's unique DNA sequences between the telomeres.

Telomeres and telomerase are especially important, not only in studies on aging but also in cancer research, so hundreds of scientists are now studying them.[7] Telomeres also provide exceptionally detailed information about how human chromosome 2 evolved from a fusion. In 1991, a group of scientists at Yale University isolated the site in the middle of human chromosome 2 where the two ancestral chromosomes fused and sequenced the DNA at that

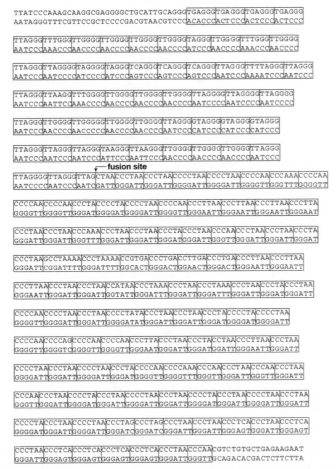

Figure 5.2. Remnants of telomeric DNA repeats at the site in human chromosome 2, where two chromosomes fused into a single chromosome about five million years ago in human evolutionary ancestry. Each boxed sequence is derived from one six-base-pair telomeric repeat. The fusion site is labeled where the DNA sequence reverses itself.

site. What they found is nothing short of spectacular. The DNA sequence at that site (shown in figure 5.2, above) has 158 copies of the six-base-pair repeat found in telomeres in the middle of chromosome 2 *right where we expect these repeats to be if human chromosome 2 arose when two ancestral chromosomes fused with each other head-to-head at their telomeres.*

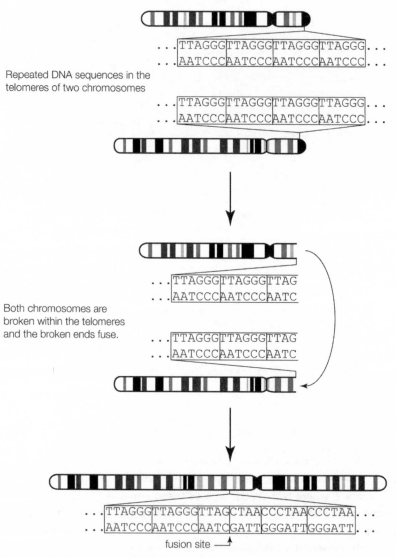

Repeated DNA sequences in the
telomeres of two chromosomes

...TTAGGGTTAGGGTTAGGGTTAGGG...
...AATCCCAATCCCAATCCCAATCCC...

...TTAGGGTTAGGGTTAGGGTTAGGG...
...AATCCCAATCCCAATCCCAATCCC...

Both chromosomes are
broken within the telomeres
and the broken ends fuse.

...TTAGGGTTAGGGTTAG
...AATCCCAATCCCAATC

...TTAGGGTTAGGGTTAG
...AATCCCAATCCCAATC

...TTAGGGTTAGGGTTAGCTAACCCTAACCCTAA...
...AATCCCAATCCCAATCGATTGGGATTGGGATT...
fusion site ⟶

Figure 5.3. How two ancient chromosomes fused at their telomeres to produce human chromosome 2.

This evidence tells us that at some point after the ancestral lineages leading to humans and chimpanzees diverged, two chromosomes fused with each other to form a single chromosome within the lineage leading to humans. In all other primates examined, these two chromosomes have remained separate. The DNA sequence shows us *exactly* where the fusion took place, right down to the individual base pairs (see "fusion site" label in figure 5.2, page 137). The fusion site is evident because the repeated telomere sequences reverse themselves at that site. As shown in figure 5.3 (page 138), this sequence reversal is precisely what a fusion of two chromosomes within their telomeres should produce.

Some of the repeats in this ancient telomere sequence are not perfect copies—they have mutated over time, although they still retain their resemblance to the original repeats. This is expected if the fusion happened long ago in human evolution. After the fusion, the telomeres at the fusion site no longer functioned as telomeres because they were no longer at the ends of the chromosome. This ancient telomere sequence, embedded in the middle of chromosome 2, had become a functionless relic of what once were functioning telomeres. Because it had entirely lost its telomere function, this sequence could accumulate mutations with no effect whatsoever; the sequence is neither beneficial nor detrimental, so it was not subject to preservation through natural selection. This relic of telomere sequence was (and is) selectively neutral.

The longer the fused telomeres persisted, the more mutations they accumulated, a common process called *mutational decay* that often happens in selectively neutral DNA. The mutational decay in this ancient telomeric sequence is high enough to indicate that the fusion must have happened in a remote hominin ancestor about five million years ago—about the time of the divergence of the human ancestral lineage from the chimpanzee ancestral lineage. Although we cannot examine their chromosomes, it is almost certain that Ardi, the Taung child, Lucy, Selam, Turkana Boy, and all other hominins from about five million years ago to the present carried this chromosome fusion.

DNA ANALYSIS SHOWS HOW INVERSIONS HAPPENED.

Let's now turn our attention to the nine inversions that distinguish human and chimpanzee chromosomes. Each of these inversions must have happened in either the human ancestral lineage or the chimpanzee ancestral lineage sometime after these two lineages diverged. The question, then, for each inversion is which species carries the original ancestral conformation and which carries the derived inversion?

There's a simple way to answer this question for each inversion. We compare the human and chimpanzee chromosomes containing these inversions with the corresponding chromosomes of other primates, such as gorillas, orangutans, and monkeys. The species—human or chimpanzee—that shares its conformation with the other primates must have the original ancestral conformation, whereas the species with the unique conformation must carry the derived inversion. Such an analysis clearly shows that the inversions in chromosomes 1 and 18 happened exclusively in the human ancestral lineage and those in chromosomes 4, 5, 9, 12, 15, 16, and 17 happened exclusively in the chimpanzee ancestral lineage.

Hildegard Kehrer-Sawatzki, a professor at the University of Ulm in Germany, led a team of researchers who identified in detail the DNA sequences

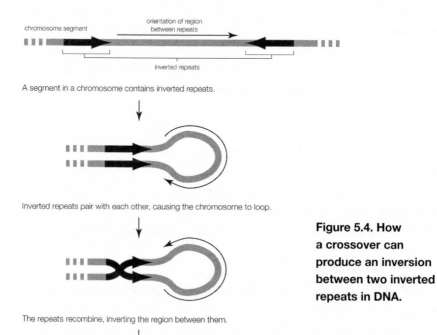

A segment in a chromosome contains inverted repeats.

Inverted repeats pair with each other, causing the chromosome to loop.

The repeats recombine, inverting the region between them.

An inversion is established in the chromosome.

Figure 5.4. How a crossover can produce an inversion between two inverted repeats in DNA.

at the ends of each inversion to reveal how each inversion happened. Let's look at two examples.

The inversion in chromosome 18 happened exclusively in the human ancestral lineage—no other living species has this inversion (although some of our extinct hominin relatives probably had it). In human chromosome 18, a segment of about 19,000 base pairs is repeated in reverse on both ends of the inversion. By contrast, chimpanzee chromosome 18 has only one copy of this repeated segment. When there are two copies of a repeated segment in reverse orientation, these repeats may pair with each other and exchange segments, as shown in figure 5.4 (page 140). This type of exchange is known as a *crossover*.

Crossovers are part of a natural and frequent process that happens during the formation of egg and sperm cells. Each of us inherits two copies of each chromosome, one from each parent. So each of us has a maternal and a paternal copy of chromosome 1, and chromosome 2, and so on. During the initial stages of egg formation in females and sperm formation in males, the two copies of each chromosome pair with each other and exchange segments through crossovers, which shuffle the DNA between the maternal and paternal copies of paired chromosomes. Just as similar chromosomes can pair and exchange segments through crossovers where they match up, repeated segments in a *single* chromosome can also pair and exchange segments to produce an inversion, although this type of pairing and crossing over is extremely rare. At some point in our evolutionary history, a segment of the DNA in chromosome 18 was duplicated, and a crossover inverted the DNA between the duplicated segments. The resulting inversion ultimately established itself in our ancestral lineage. The inversion did not happen in chimpanzee chromosome 18 because the corresponding segment of DNA is not duplicated in chimpanzee chromosome 18. It duplicated itself exclusively in the human ancestral lineage after it had diverged from the chimpanzee ancestral lineage.[8]

A second example is the inversion in chimpanzee chromosome 9. For reasons that are not fully understood, this chromosome is especially prone to inversions. In modern humans, new inversions arise more often in chromosome 9 than in any of the other chromosomes, and these inversions are a common cause of reduced fertility in humans alive today.[9] The same is true for primates in general—new inversions appear in this same chromosome in modern great apes and monkeys at an unusually high rate, just as they do in humans.[10]

Beyond new inversions, there is good evidence that this chromosome has been accumulating inversions throughout its evolutionary history. Humans and all great apes carry an ancient inversion in this chromosome that happened more than ten million years ago. Then, after the ancestral lineage leading to gorillas split from the one leading to humans and chimpanzees, yet another inversion happened in the common ancestor of humans and chimpanzees, and it is now present in all humans, chimpanzees, and bonobos. Finally, another inversion took place after the human-chimpanzee ancestral divergence, and it is present only in common chimpanzees and bonobos. It arose from a complex set of multiple rearrangements, including one large inversion accompanied by a smaller inversion.[11] Chromosome 9 was and is clearly prone to inversions.

IT IS POSSIBLE TO RECONSTRUCT THE EVOLUTIONARY HISTORY OF CHROMOSOME REARRANGEMENTS.

Let's briefly review what we already know about human and chimpanzee chromosomes and their rearrangements. The modern human genome contains twenty-three chromosomes, and the chimpanzee genome has twenty-four (humans have one fewer chromosome than chimpanzees because of the chromosome-2 fusion). The inversions in human chromosomes 1 and 18 and the fusion in chromosome 2 arose in the human ancestral lineage, whereas the inversions in chimpanzee chromosomes 4, 5, 9, 12, 15, 16, and 17 arose in the chimpanzee ancestral lineage. From this information, we can reconstruct the chromosome configurations in the common ancestor of humans and chimpanzees. That ancestral species had twenty-four chromosomes in its genome, four of which had the same conformation as chimpanzee chromosomes 1, 18, and 2A and 2B (the two chimpanzee chromosomes corresponding to human chromosome 2). Seven of the chromosomes had the same conformation as human chromosomes 4, 5, 9, 12, 15, 16, and 17. The remaining thirteen chromosomes were the same as their counterparts in both humans and chimpanzees.

By comparing chromosome rearrangements across multiple species, scientists can extend this type of analysis back through evolutionary time to tell how and approximately when chromosome rearrangements happened and then reconstruct how the chromosomes evolved from common ancestors. We now have a good idea of what the chromosome conformation was in the common ancestor of all modern primates. This ancestor lived about eighty-five to ninety million years ago, and it had twenty-five chromosomes in its genome, two more than humans and one more than modern great apes.[12]

Compared to the chromosomes of this common ancestor of living primates, human chromosomes contain a plethora of rearrangements—evidence that chromosome rearrangements repeatedly accumulate over evolutionary time. Most are inversions and translocations (exchanges of chromosome segments between two different chromosomes). Fusions and fissions (breakage of one chromosome to form two) also happened. Human chromosomes 2, 7, 10, and 16 arose from fusions of ancient chromosomes at different points in our evolutionary history (chromosome 2 being the most recent, as we have seen). And two ancient chromosomes underwent fissions, resulting in what are now human chromosomes 3, 14, 15, and 21.

ARE NEW CHROMOSOME REARRANGEMENTS STILL APPEARING IN HUMANS?

The answer to this question is an unequivocal *yes*. Newly arisen inversions, translocations, fusions, and fissions repeatedly happen in humans as well as in many other species. Some are associated with genetic disorders. A small proportion of people with Down syndrome have the condition because of a fusion between chromosome 21 and either chromosome 14 or 15. Also, an inversion in the X chromosome is the most common mutation responsible for hemophilia. Most people who carry chromosome rearrangements have no outward symptoms, but some may have a reduction in fertility. And it is this effect—*reduction in fertility*—that underscores the crucial role of chromosome rearrangements in evolution.

As mentioned earlier, we inherit two copies of each chromosome—one from

each of our parents. During the formation of egg and sperm cells, these two
chromosomes align with each other and exchange segments through crossovers.

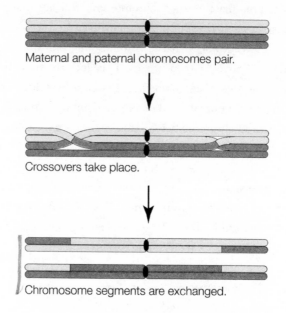

Maternal and paternal chromosomes pair.

Crossovers take place.

Chromosome segments are exchanged.

Crossing over is one way that nature recombines genetic material, creating new combinations of genes. Without it, variations in DNA on the same
chromosome would remain inextricably linked, but because of crossing over,
new combinations of genes in the same chromosome are possible.

If someone inherits a chromosome with an inversion from one parent and
a copy of that same chromosome but without the inversion from the other
parent, that person is said to be *heterozygous* for the inversion. But when it
comes time for the chromosomes to align and pair, the inversion in one chromosome cannot directly align with its corresponding noninverted segment in
the other chromosome.

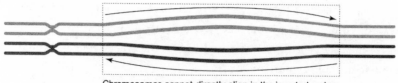

Chromosomes cannot directly align in the inverted region.

Instead, to overcome this nonalignment, the chromosomes loop in what could be called an alpha-omega loop—one chromosome loops like a lower-case Greek alpha (α) tilted ninety degrees to the right, while the other loops like an uppercase Greek omega (Ω), allowing the two chromosomes to align throughout their full lengths.

With the chromosomes fully paired in their looped conformation, all is fine and good—unless a crossover happens within the looped region, which ends up deleting large segments and duplicating other large segments, so that the resulting chromosomes have large portions of missing and extra genetic material. In genetic terms, we describe cells that inherit these chromosomes with deleted and duplicated segments as *unbalanced*. Think of them simply as cells that have so much missing and extra material that they can no longer function as they should.

When such crossovers occur in a person who is heterozygous for an inversion, some of the egg or sperm cells are unbalanced. If these unbalanced egg or sperm cells undergo fertilization, the resulting embryo is also unbalanced and, in most cases, is spontaneously aborted (miscarried). The result is reduced fertility for any person who is heterozygous for an inversion. The same is true for people who are heterozygous for other chromosome rearrangements, such as translocations or fusions. They also suffer a reduction in fertility because a portion of their egg or sperm cells are unbalanced.

Not surprisingly, chromosome rearrangements are a common cause of infertility in humans. When people who suffer from infertility seek medical attention, a fertility specialist may examine the patient's giemsa-stained chromosomes to determine whether he or she is heterozygous for a chromosome rearrangement.

CHROMOSOME REARRANGEMENTS CONTRIBUTE TO THE EMERGENCE OF NEW SPECIES.

Inversions and other chromosome rearrangements cause a loss of fertility in people who are heterozygous, so shouldn't natural selection eliminate these rearrangements? A single rearrangement typically causes a *reduction* in fertility but not a complete *loss* of fertility for individuals who are heterozygous for the rearrangement, and such individuals tend to have fewer offspring. But individuals who inherit the *same* chromosome rearrangement from both parents are said to be *homozygous* for the rearrangement, and their chromosomes align and pair normally. They carry the rearrangement in *both* copies of the chromosome and have no reduction in fertility. There is no looping, no unbalanced cells, and no reduction in fertility.

Only individuals who are *heterozygous* for a rearrangement suffer a reduction in fertility, which explains why chromosome rearrangements are important in evolution. Natural selection favors those individuals who survive *and reproduce*. Therefore, natural selection favors those who are homozygous for a chromosome rearrangement, or homozygous for the original conformation, and *disfavors* those who are heterozygous because of their reduction in fertility.

In nature, when groups of individuals are isolated for long periods of time, such as fish in different lakes or river systems, or animals separated by natural barriers like mountain ranges or oceans, we expect chromosome rearrangements to accumulate differently over time in the two groups, just as mutations accumulate differently. Both genetic theory and experimental observation show that, over many generations, a rearrangement ultimately becomes homozygous in all the individuals of one group while another group may remain homozygous for the original conformation. This explains why all humans are homozygous for the fused chromosome 2 and for inversions in chromosomes 1 and 18. At different points in our ancient ancestry, each of these rearrangements happened in one individual, each of whom was a common ancestor of all people alive today. We might call them the chromosome 1, 2, and 18 Eves or Adams, much like the mitochondrial Eve or Y-chromosome Adam, although in the case of chromosomes 1, 2, and 18, there's no way to determine whether each common ancestor was female or male. Over many generations, these chromosome rearrangements became homozygous in the human ancestral lineage, and the original conformations were lost in that lineage. The original conformations of these chromosomes, however, remained homozygous in the lineages leading to all other primates.

Over vast amounts of time, different chromosomal rearrangements have accumulated in groups that at one time had common ancestry but were later isolated. When individuals of two long-isolated groups are brought back together, they may be so different genetically that they are no longer interfertile; in other words, no offspring can be produced from a hybrid mating. They have evolved to become fully separate species. On the other hand, if the genetic changes are not too drastic, hybrid offspring may be produced, but the differences for several chromosome rearrangements render them partially or fully infertile.

This is what happens when a horse and donkey mate to produce a mule. Horses and donkeys differ by several chromosome rearrangements, each of which is now homozygous in all members of its respective species. Mules are heterozygous for all these rearrangements, and their chromosomes are beset with so many loops and haphazard arrangements that they cannot properly align and sort into egg and sperm cells. The result is nearly complete infertility.

The infertility of interspecific hybrids raises an almost unthinkable question, but it is one that often arises when discussing the evolutionary history of our chromosomes. Is our evolutionary relationship to chimpanzees still close enough for infertile hybrids to be produced? Speculation and hype about human-chimpanzee hybrids has been popularized for more than a century. One of the most recent cases was an upright-walking chimpanzee named Oliver, who had been raised by people as a pet since infancy. From the 1970s through the 1990s, speculation about his possible human-chimpanzee hybrid origin abounded. He was paraded as a spectacle onstage and became the subject of television documentaries, one as recent as 2006. Genetic testing in the 1990s, including giemsa staining of his chromosomes, however, showed that he was a pure common chimpanzee, with DNA sequences consistent with an origin in central Africa, probably in Gabon.[13]

Most people today (myself included) find the idea of human-chimpanzee hybridization repulsive, and if it were attempted, it would be a severe breach of scientific ethical conduct and the laws of most countries. Given current human-rights laws and scientific standards governing research on human and animal subjects, no reputable scientist would consider conducting an experiment to test this question, nor would any legitimate scientific institution allow such an experiment.

But times were different in the early twentieth century, when wild speculation sometimes displaced reliable science. In a racist interpretation of human evolution, the anthropologist Hermann Klaatsch speculated (with no scientific evidence) that people of Asian origin had evolved from orangutans—those of African origin from gorillas, and those of European origin from chimpanzees.[14] Believing this to be true, an American lawyer, Howell S. England, attempted to raise money for hybridization experiments between people of different ethnic backgrounds with their purported ancestral ape species.[15]

This time was also the age of eugenics, a movement that promoted development of a so-called superior human society by forcibly sterilizing people who were considered to be genetically inferior. In the United States, various states passed eugenics laws that prohibited interracial marriage and required mandatory sterilization of people identified as genetically inferior, which often included the poor and uneducated and those who were diagnosed as being feeble-minded. More than sixty thousand people in the United States were involuntarily sterilized under these laws.

During this period, Ilya Ivanovitch Ivanov, a Soviet zoologist, proposed conducting human-ape hybridization experiments.[16] Ivanov was an expert in artificial insemination and had applied his expertise to revolutionize horse breeding throughout the Soviet Union. In 1924, he submitted a request for funding artificial insemination between humans and apes, and the Soviet government approved it in 1925. Shortly thereafter, Ivanov departed for French Guinea to conduct his experiments. He had planned to artificially inseminate native women with chimpanzee semen without their consent. However, the colonial governor strongly objected, prompting Ivanov to appeal to his superiors in Moscow. After deliberation, a committee, which had been convened to consider his request, forbade him from artificially inseminating women without their consent.

Ivanov turned instead to female chimpanzees. He artificially inseminated three female chimpanzees with human sperm, but all three failed to become pregnant. He left French Guinea with twenty chimpanzees, intending to transport them to the Soviet Union for further experiments. Sixteen of them died during the journey.

The Soviet government established a primate research facility where Ivanov could continue his work. After a lengthy review, his request to artificially inseminate human female volunteers with ape semen was approved in 1929. Five women volunteered to be the subjects of the experiment, and they were to raise their hybrid children at the facility. However, by then, the last male ape at the facility who could donate semen, an orangutan, had died. Another shipment of apes arrived at the facility in 1930, but before Ivanov could initiate the experiments, he was arrested and imprisoned as part of a political purge of scientists mandated by Joseph Stalin. Ivanov died in prison in 1932, and his proposed experiments on human-ape hybridization were abandoned.

Though largely forgotten in later years, Ivanov's experiments were sensationalized during their time. They came on the heels of the 1925 Scopes trial, a media sensation focused on teaching evolution in Tennessee and the first trial to be broadcast live on radio throughout the United States. The trial had already ignited passions and divided the American public into separate camps—those who embraced evolution and those who disdained it. Publicity of Ivanov's work the following year added fuel to the fire and led some to connect evolution with communism, a notion that persisted for decades thereafter—and still persists in some people's minds.

As political turmoil erupted and World War II ensued, eugenics efforts reached their horrific climax with the Nazi Holocaust. A lesser-known Nazi effort was the Lebensborn program, in which women identified as being "racially superior" were to bear the children of SS soldiers to produce a so-called genetically superior race. At the end of the war, as the atrocities of Nazi eugenics programs were publicized, sentiment against eugenics began to grow. The once-popular idea of hybridizing humans and apes was ultimately viewed as repulsive, and Ivanov's experiments were largely forgotten—documentation of them was kept hidden in the Soviet Union. Only recently have the original documents resurfaced, allowing the detailed story to be told.

Given what we know about human genetics, it is highly unlikely that Ivanov's experiments could have produced a human-ape hybrid. Chromosome rearrangements are only one mechanism contributing to the reproductive isolation of humans and chimpanzees; variations in genes that govern reproductive processes, along with a host of other factors, combine to reproductively isolate humans from our evolutionary relatives. For example, some of the most highly diverged genes in humans and chimpanzees are those that govern the formation and compatibility of egg and sperm cells.[17] Moreover, infertile hybrids are usually possible only when two *very* closely related species are hybridized. The evolutionary relationship between horses and donkeys, for example, is much closer than the relationship between humans and chimpanzees. The most recent common ancestor of horses and donkeys lived about two million years ago, whereas the most recent common ancestor of humans and chimpanzees dates to a much more distant time, about five to seven million years ago.

Some nonhuman primate species are much more closely related than humans and chimpanzees, such as the common chimpanzee and the bonobo. However, no case of common chimpanzee–bonobo hybrid has ever been scientifically documented. Common chimpanzees and bonobos have the same inversions on chromosomes 4, 5, 9, 12, 15, 16, and 17, but a few smaller chromosome rearrangements distinguish the two species.[18]

Not long ago in evolutionary time, humans and Neanderthals coexisted as closely related species. As we saw in chapter 3, there is evidence of limited human-Neanderthal and human-Denisovan hybridization outside of Africa.[19] The divergence of the lineages leading to humans and the common ancestor of Neanderthals and Denisovans about eight hundred thousand years ago was apparently recent enough for hybridization and for fertility to persist. Analysis of the Neanderthal genome may soon reveal which chromosome rearrangements they shared with us, although that information is not yet available.

We've looked at how major rearrangements of genomes played a crucial role in human evolution and in the evolution of other species. In the next chapter, we'll focus on the most important part of our genome—the DNA in our genes—to read the story it tells us about our evolutionary history.

CHAPTER 6
MORE EVIDENCE FROM OUR GENOME
THE GENES THAT MAKE US HUMAN

For eight years, Gregor Mendel cultivated thousands of pea plants in his monastery garden and greenhouse, hybridizing them in various combinations and meticulously recording the numbers of their progeny with variations in inherited characteristics. His work was unusual for nineteenth-century biology in its size and scope. In the end, Mendel derived the fundamental principles of inheritance, which years after his death would earn him recognition as the founder of genetics. Arguably, he and Darwin were the two nineteenth-century scientists who would most influence the future of biology. The synthesis of their discoveries would eventually establish the modern theory of evolution.

Unlike Darwin, Mendel was right about almost everything he addressed; his experiments, and his interpretations of them, stand as valid today as they did when he published them in 1866. Lamentably, as we saw in the first chapter of this book, Mendel failed to enjoy recognition of his work during his lifetime. No one understood the importance of what he had discovered, and his work lay in near obscurity for thirty-five years, sixteen of them after his death. Although he was scientifically active at the same time as Darwin, the two great scientists never met or corresponded, and nearly seventy years would pass before their discoveries and ideas were fully integrated.

Mendel deduced how simple contrasting variations in pea plants, such as differences in flower color or seed shape, were inherited. And he recognized how his discoveries could explain evolution. In doing so, he firmly established

151

what a gene is at the level of inheritance, although he never used the word *gene* (which had not yet been coined), instead referring to hereditary factors or elements. Unfortunately, even today, biology students often draw an incorrect conclusion about genes when studying Mendel's experiments. They presume that, according to Mendel, one gene confers one trait, something that Mendel never intended to convey. Statements such as "the gene for seed shape" in Mendel's experiments or "the gene for eye color" when referring to humans propagate this misconception of one gene for one trait.

This misconception can have serious consequences, which became painfully obvious to me several years ago, as I was watching an episode of a popular investigative news show on a major US television network. It told the story of a divorced man who picked up his son from his ex-wife's home while he was with a woman whom he was dating. The woman noticed that both the man and his ex-wife had blue eyes and their son had brown eyes. According to her (mis)understanding of Mendelian genetics, she presumed that it was impossible for two blue-eyed parents to have a brown-eyed child, so she determined that the man could not be the biological father of the boy and told him so. At that point in the story, the television narrator's voice interjected as an image of an open biology textbook appeared. The voice said something like this: "The gene for blue eyes is recessive. Mendel's laws prove that two recessives can't make a dominant, so it's impossible for two blue-eyed parents to have a brown-eyed child. The boy could not be his son."

I pulled off my shoe and threw it at the television screen (this was in the days of glass-screen TVs, so my shoe did no damage). My reaction came from the thought of thousands of brown-eyed people across the United States whose parents had blue eyes needlessly questioning their paternity. In this particular case, a DNA test (which *is* reliable) confirmed the woman's suspicion—the boy was *not* the man's biological son. However, purely on the basis of eye color, the narrator and the woman had it wrong. There is no such thing as *the* gene for eye color. Rather, there are *several* genes that influence eye color, and they interact with one another to give us the wide range of eye colors we see in people.

It is true that many cases of blue eyes are largely a consequence of *variation* in one of the several genes that influence eye color, and that the variant responsible for blue eyes is recessive. Therefore, variation in a single gene can result in blue eyes. However, because more than one gene may be involved in producing eye color variation, some blue-eyed parents *can* have brown-eyed children. Notably, one of the world's leading human geneticists, Victor McKusick, had brown eyes, along with his identical twin brother, and both of their parents had blue eyes.[1]

The notion of *the* gene for this trait or *the* gene for that trait implies a one-to-one relationship of genes and traits—one gene for each trait. But this is a common yet misleading way to think of genes. What is a gene? Simply put, a gene is a segment of DNA that encodes an RNA molecule, and the RNA molecules encoded by most genes, in turn, encode proteins. This concept of information transfer is known as the central concept of molecular biology, and it can be represented as

$$DNA \longrightarrow RNA \longrightarrow protein$$

where the arrows represent a transfer of information, not a chemical conversion. Each protein encoded by a gene carries out a specific function, which, coupled with the functions of other proteins encoded by other genes, ultimately confers a trait, such as eye color.

We now know much about how genes determine eye color, and this process turns out to be a complex interaction of proteins encoded by several genes. Some of the proteins convert one substance into another, some activate or suppress other genes, and some regulate the cell environment, such as controlling the degree of acidity when the pigment is being made.

As a simple but effective analogy, we can think of genes as the pages of sheet music read by members of a symphony orchestra. The pages make no music themselves, but they contain the directions the musicians follow. Symbols on the pages direct some to play at one point while others remain silent. At particular times, certain members of the orchestra play loudly, other times softly, all according to the musical score. Over the course of the symphony, the music rises and subsides, with each orchestra member playing a predetermined part at the right time and in the right way, blending her or his playing with all the other players, all as directed by the musical score. In the same manner, genes are turned on and off in the cells of the body, like players in a biological symphony, some producing their proteins, others remaining silent, all in the right way and at the right time in a coordinated fashion that over time guides development and function of the human body throughout a lifetime. Our genes even play a role in aging and ultimately death from old age, if we are fortunate enough to escape death from some other cause.

HOW MANY GENES DO WE HAVE?

When I was an undergraduate student in the early 1980s, most scientists thought that the human genome contained about 200,000 genes. By the time I was a graduate student in the late '80s, the estimate was half that: about 100,000. When the first draft of the human genome's DNA sequence was released in 2001, the number had dwindled to about 35,000 genes. But even that estimate was too high. When a highly finished version of the human genome was released in 2004, the number of known genes was confidently placed at 19,438, with an additional 2,188 segments of DNA that had all the characteristics of genes but had not yet been confirmed as true genes.[2] In other words, the number of genes we have is somewhere around 20,000.

If the DNA in any one of those genes mutates, the protein encoded by that gene can be altered, leading to possible changes in any of the traits influenced by that gene. If the DNA changes ever so slightly by mutation in several genes, we see a wide range of variation, such as variation for pigmentation in eyes, hair, and skin; differences in height, weight, and face shape; and all the many inherited variations that make possible the wonderful genetic diversity in our worldwide human family.

HOW DO OUR GENES EVOLVE?

A key component of evolution is the modification of inherited characteristics in a species. And to be inherited, those outward modifications are determined by modifications in the sequence of DNA. Therefore, our DNA can reveal how evolution transpired at its most fundamental level. Clues from the human genome, as well as from genomes of other species, offer clear evidence of how our genes evolved through mutation—and how they continue to evolve.

As it turns out, there are several pathways for gene evolution, and we'll look at the evidence of a few of them here. In the simplest case, a single gene acquires mutations that give it a beneficial function. Because the mutated gene is more beneficial than the original version, the individuals who carry it, and their offspring who carry it, are favored through Darwinian natural selection. Eventually, over a period of generations, all members of the species ultimately inherit the mutated gene in place of it its previous version. This process of mutation followed by natural selection favoring new mutations is called *positive selection*. When it happens across many genes over many generations, it contributes to the evolution of new species.

Some opponents of evolution argue that, according to modern science, mutations are never beneficial, so positive selection is a myth promulgated by evolutionary theoreticians.[3] This notion is easily discredited with evidence from the human genome, and we'll look at some examples of positive selection momentarily. However, before doing so, we must make one point right up front—the vast majority of mutations are *not* beneficial. Most have essentially no effect—they are neither beneficial nor harmful—and natural selection has little influence on them. Scientists call them *selectively neutral mutations*.

Most mutations that *do* have an effect are harmful. They often disable genes by disrupting their normal function. Examples of harmful mutations are numerous, and they have been documented in large databases. They are responsible for a multitude of human genetic disorders, among them sickle-cell anemia, cystic fibrosis, phenylketonuria (PKU), Huntington disease, and fragile X syndrome, to name a few. In most cases, harmful mutations are gradually purged from a species as individuals who display the harmful trait associated with the mutation have fewer offspring than those who don't. This purging process is called *negative purifying selection*, and it is the most common form of natural selection—"negative" because it *disfavors* new mutations, and "purifying" because it tends to *preserve* the existing genetic situation. When spread over the genes of a species, it stabilizes the species, slowing evolutionary change. There is abundant evidence of it in all species studied, including humans.

Many of these harmful mutations that cause genetic disorders are what we call *recessive mutations*, which means that, for someone to have the genetic disorder, she or he must inherit mutated copies of the gene from both parents. If someone inherits a mutated copy of the gene from one parent but not the other, that person shows no symptoms of the genetic disorder. He or she is said to be a *carrier* of the mutated gene, and, on average, half of that person's children inherit the mutated gene and are likewise carriers. The mutated gene may persist undetected, passed on from one carrier to another for many generations until the two parents of a child happen to both be carriers. If any child inherits mutated copies of the gene from *both* parents, that child will have the genetic disorder. We all are carriers of recessive mutations that remain silent within us, their effects appearing only if our descendants happen to inherit them from both parents.

Over long periods of evolutionary time, these harmful mutations gradually disappear though negative purifying selection. However, because natural selection has no effect on carriers of recessive mutations, it is very ineffective at entirely eliminating recessive mutations. Elimination will eventually take place, but it can require many generations to do so. By then, new recessive mutations in the same gene may have appeared elsewhere in the species, so a genetic disorder associated with recessive mutations in a gene may never entirely disappear.

An example is hemophilia, a genetic disorder that prevents blood clotting and causes people who have it to bleed excessively when bruised or cut. Until recently, many people with hemophilia suffered premature death from bleeding, but now, genetically engineered proteins effectively treat hemophilia, allowing people with hemophilia to lead nearly symptom-free lives. Because the disorder often caused premature death throughout most of human history, people with hemophilia typically had either no children or fewer children than their relatives who did not have hemophilia. Consequently, negative purifying selection tends to reduce and ultimately eliminate mutated versions of the gene in humans.

Hemophilia among Queen Victoria's descendants is an example. She was a carrier of a new recessive mutation that probably took place in her father and was transmitted to her in the sperm cell that contributed his chromosomes to her. It's often said that the several cases of hemophilia in Queen Victoria's descendants were a consequence of *inbreeding*—mating between close relatives.

Although inbreeding is well documented in European royal families, speculation that it contributed to the incidence of hemophilia in Queen Victoria's
descendants is unfounded. The mutation is in a gene on the X chromosome.
Each female inherits two X chromosomes, one from her mother and one from
her father, and so she can be a carrier of a recessive mutated gene without
having hemophilia. Hemophilia can be attributed to inbreeding only when a
female has hemophilia because she inherited the same mutated gene from *both*
parents due to their recent common ancestry, and none of Queen Victoria's
female descendants had hemophilia. But a male inherits only one X chromosome—from his mother (the father contributes a Y chromosome). Any male
who inherits the mutated gene on his X chromosome has hemophilia. So, con-
trary to popular legend, inbreeding had nothing to do with the incidence of
hemophilia in European royal families.

Of Queen Victoria's nine children, one had hemophilia (Leopold), and two
were carriers (Alice and Beatrice). All three passed the mutated gene to some
of their children, and for three generations hemophilia appeared in several of
Queen Victoria's male descendants. These males with hemophilia had relatively few children for a variety of reasons, some of them political but also in
some cases because of premature death due to excessive bleeding. Thus negative purifying selection (in this case, the premature death of some males as a
consequence of their hemophilia) was one, albeit not the only, factor contributing to the decline and ultimate disappearance of this mutated gene. Current
evidence indicates that the mutated gene probably did not persist beyond the
third generation. However, other mutations in this same gene have caused
hemophilia in other families, so the disorder remains in the human species as
new mutations arise, while others are gradually eliminated.

Although neutral and harmful mutations are more common than beneficial ones, there are numerous examples of beneficial mutations that natural
selection has favored, resulting in progressive change within a species and
ultimately contributing to divergence of species. One of the most intriguing
examples is the *FOXP2* gene in humans. This gene made headlines in 2001
when a group of scientists at Oxford University found that mutations in this
gene can cause severe speech disorders, implying that this gene is essential for
human speech.[4] Because articulate speech is unique to our species, scientists

wondered if mutations in this gene were, at least in part, responsible for the evolution of speech.

As it turns out, FOXP2 is one of many important genes that regulate the development of brain and nerve cells in mammals. When scientists compared the human version of the FOXP2 gene with the corresponding gene in a wide range of mammalian species, they found two mutations in the human version that caused important alterations in the protein encoded by this gene.[5] One of these mutations appears to be unique to humans and has not been found in any other living species, although there is evidence that Neanderthals carried it.[6] The protein encoded by FOXP2 functions differently in humans than in other species, and it probably regulates a suite of genes that contribute to the human brain's ability to control speech.

When positive selection strongly favors a beneficial mutation, the mutation can rapidly spread throughout the species over a small number of generations until the original version of the gene is entirely displaced and every member of the species carries the mutated gene. This phenomenon is called a *selective sweep*, and it leaves a distinct imprint on the DNA when it happens. If the progress of natural selection favoring a mutated gene is slow, requiring many generations over a long period of time, neutral mutations located near the favored mutation in the DNA can gradually be separated from it. However, when there is a rapid selective sweep, these nearby mutations tend to "piggyback" along with the favored mutation. The DNA in and near the FOXP2 gene in humans bears this pattern of a rapid selective sweep, evidence that at least one mutation in the human FOXP2 gene conferred a powerful advantage on ancient humans.[7] This example is one of several in which a mutation altered a gene product, and the alteration gave the gene's product a new and beneficial function, which was then favored through natural selection.

EVEN DISABLING MUTATIONS MAY BE FAVORED THROUGH NATURAL SELECTION.

We're now going to explore one of the most intriguing cases of evolution in human history, one that highlights not just our own evolution but how our evolution is inextricably tied to that of other species. Malaria is one of the most horrific of human diseases. It afflicts as many as 250 million people, most of them in tropical regions of the world where mosquitoes are common. There are several forms of malaria, but the most widespread and severe type is caused by a microorganism called *Plasmodium falciparum*. It cannot be transmitted directly from person to person but must be carried through mosquitoes, acquired and transmitted when a mosquito bites a person.

Chimpanzees also suffer from a less severe form of malaria caused by a related microorganism, *Plasmodium reichenowi*. For many years, scientists suspected that the modern human malarial parasite *Plasmodium falciparum* and the modern chimpanzee parasite *Plasmodium reichenowi* diverged from a common ancestral parasite about five to seven million years ago along with divergence of the ancestral lineages that led to their host species—humans and chimpanzees. However, a discovery published in 2009 told a much different story. A team led by Stephen Rich at the University of Massachusetts–Amherst, Francisco Ayala at the University of California–Irvine, and Nathan Wolfe of the Global Viral Forecasting Initiative made a startling discovery. The human parasite *Plasmodium falciparum* had evolved from the chimpanzee parasite *Plasmodium reichenowi* by jumping hosts from chimpanzees to humans as recently as five to ten thousand years ago.[8]

Evidence in the human and chimpanzee genomes showed how it happened. When *Plasmodium reichenowi* infects chimpanzees, it recognizes a substance on the red blood cells that arises from a chimpanzee gene called *CMAH*. We have this same gene, *CMAH*, and it is in the same place in our DNA as in that of chimpanzees. However, our version of *CMAH* is mutated so that it no longer functions. It has become what we call a *pseudogene*—a disabled gene. And no one has the original non-mutated version—it was lost long ago in our ancestry. As a consequence, we do not make this substance as chimpanzees do, although our ancient ancestors once did.

The mutation that disabled this gene happened more than two million years ago, when all hominins were still exclusively in Africa where malarial parasites were and are endemic. And after the mutations disabled the gene, those who inherited only the mutated form had a distinct advantage over those who carried the original non-mutated gene—they were resistant to malaria.

There is evidence in our DNA of a selective sweep favoring this mutated *CMAH* pseudogene in ancient humans.[9] Those who inherited it from both parents were resistant to the type of malaria prevalent at the time. They survived and had more progeny than those who had a copy of the original gene and were susceptible to the disease. Through this selective sweep, hominins with the mutated *CMAH* pseudogene survived and reproduced so successfully that the pseudogene became the only form of this gene in our ancestral lineage. Eventually, everyone had only the mutated pseudogene, and everyone was resistant to malaria. And that situation persisted in our ancestors for more than two million years. Interestingly, Neanderthals apparently carried this same *CMAH* pseudogene, evidence that the mutation and selective sweep preceded the divergence of the lineages leading to humans and the African ancestors of Neanderthals.[10]

Parasites are also subject to mutation and natural selection, which often allows them to overcome natural resistance. When the malarial parasite overcame human resistance, the result was devastating. Mutations in a gene called *EBA-175* in *Plasmodium reichenowi* allowed those parasites that carried the mutation to recognize another substance, which is abundant on human red blood cells. About five thousand to ten thousand years ago, the final mutations arose, allowing the newly evolved parasite to infect humans. The parasite jumped from chimpanzees to humans, and *Plasmodium falciparum* evolved as a new species. We are now highly susceptible to this new human-specific form of malaria, which is far worse than the old version that once afflicted our ancestors and still afflicts chimpanzees.

Malaria is now one of the worst modern human diseases. Massive efforts to control the disease, and the mosquitoes that transmit it, have successfully reduced malarial cases worldwide. Even so, about 247 million people contract malaria, and nearly a million of them die from it every year.[11]

In this example, we see two cases of mutations in genes favored by positive selection. In one case—the *CMAH* gene in humans—a mutation that disabled the gene was beneficial to humans, conferring resistance to malaria. In the second case—the *EBA-175* gene in the parasite—a mutation that altered the gene's original function was beneficial for the parasite.

NEW GENES MAY EVOLVE THROUGH DUPLICATION FOLLOWED BY DIVERGENCE.

Human genes are typically large and complex, not the sorts of things that arise quickly. Rather, they develop through a long evolutionary process of mutation coupled with DNA expansion and rearrangement. There are a few known cases of new human genes arising anew with no derivation from previous genes, and we'll look at some examples later in this chapter. However, the vast majority of our genes evolved from preexisting genes. And the most common evolutionary process for new genes to evolve from preexisting genes is gene duplication and divergence.

Our genes are often grouped in clusters of similar genes, evidence that they arose from a single ancestral gene that duplicated itself to produce two side-by-side copies. Later, one of those copies duplicated itself to produce yet another copy, and so on until there were several copies of the gene in a cluster.

When a gene is duplicated, the constraints of negative purifying selection are relaxed. Mutations in one copy of the gene are no longer detrimental because the other copy still carries out the original function. The mutated copy can acquire an altered function, or it may even lose its function, with no ill

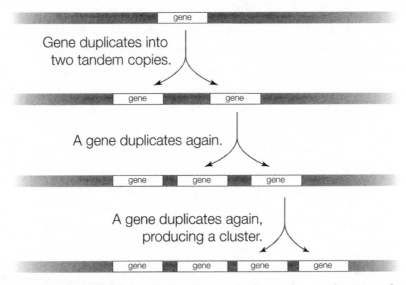

Figure 6.1. Repeated gene duplication producing a cluster of gene copies.

effects because there is at least one backup copy. Occasionally mutations in one of the gene copies confer a new beneficial function that differs from the original function. Now the mutated gene is favored through natural selection, and eventually all members of a species inherit it, along with the original gene. These duplicated genes that mutate to carry out new beneficial functions contribute to the evolution of new species by expanding the possible functions previously served by a single gene.

There are many examples of side-by-side copies of genes organized into clusters where gene duplication and divergence, coupled with natural selection, explain the various functions of genes within each cluster. More than one-third of our genes reside in such clusters, evidence that duplication followed by divergence is a major pathway for gene evolution. Even some genes that do not reside in clusters probably arose in this way, but the duplication and divergence happened so long ago in evolutionary history that DNA rearrangements have broken up the clusters.

Perhaps the best-studied example of how gene clusters evolved is the human globin clusters depicted in figure 6.2, below. These two clusters contain genes that encode the alpha and beta globins, which are proteins in hemoglobin, the substance that carries oxygen in the blood and makes our blood red. The beta-globin cluster offers an excellent example of how positive selection has favored mutations in certain genes belonging to this cluster. When a woman is pregnant, the fetus receives all its oxygen from the mother. The fetus's and the mother's circulatory systems are not interconnected, so their blood supplies are completely separate. Oxygen must diffuse from the mother's blood to the fetus's blood where their blood vessels come close to

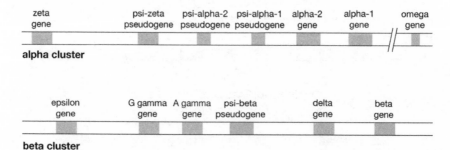

Figure 6.2. The human alpha- and beta-globin gene clusters.

each other in the placenta. Something must ensure that the oxygen flows in the right direction—from mother to fetus—otherwise the fetus would suffocate from oxygen starvation. As it turns out, there are several mechanisms that work together to ensure that the fetus obtains enough oxygen.

The gene labeled "beta gene" in figure 6.2 (page 162), provides nearly all the beta globin in adult human blood. But in a fetus, the beta gene is shut off and the two gamma genes, G-gamma and A-gamma, are turned on. The gamma genes carry mutations that slightly change the proteins they encode so that these proteins attract oxygen more strongly than does the adult beta protein. When fetal blood carrying the gamma globins comes close to the mother's blood carrying beta globin, the gamma globins attract the oxygen away from the beta globin, and oxygen flows from mother to fetus.

Is there some way to determine how such a favorable situation evolved? Fortunately, the DNA of the beta-globin gene cluster has been sequenced in a wide variety of species, and these sequences have allowed scientists to reconstruct how the gene cluster evolved. First, the genes in both the alpha and beta clusters arose from a single gene that began duplicating and diverging about 450 million years ago, when our remote ancestors were fish living in the oceans. These genes have continued to duplicate and diverge to the point that humans and chimpanzees now have thirteen copies of them: seven copies in the alpha cluster and six in the beta cluster (see figure 6.2, page 162). Evidence that the duplication process is still under way comes from the discovery that a few people have fourteen instead of thirteen copies, a consequence of a recent duplication.[12]

Two copies are evolutionary newcomers: the G-gamma and A-gamma copies in the beta-gene cluster, which encode proteins conferring high oxygen affinity to fetal blood. They are unique to primates, having arisen by duplication after primates diverged from other mammals. Originally, there was just one gamma copy, and many primates, such as lemurs and New World monkeys, still have only that single copy. The gene duplicated itself in a common ancestor of humans, great apes, and Old World monkeys, and this common ancestor lived in Africa more than twenty-five million years ago. So we now have two side-by-side copies of the gene: G-gamma and A-gamma. Over time, mutations have been favored through natural selection in the gamma genes, conferring to their proteins a strong affinity for oxygen. This affinity, in turn, permits fetuses to extract oxygen by diffusion from their mother's blood. These mutations in the G-gamma and A-gamma genes are yet additional examples of beneficial mutations favored through positive selection.

PSEUDOGENES MAY ARISE FROM DUPLICATED GENES.

Is there something that directs duplicated genes to mutate in just the right way so that they carry out new beneficial functions? A core concept of evolutionary theory is that mutations are not directed. Rather, they happen at different places in genes in different individuals, and those individuals who carry beneficial mutations are more likely to survive and reproduce than those who don't, thus passing on the beneficial mutations to offspring.

A common misperception, however, is that *undirected* means *random*. Mutations are not entirely random. Certain regions of our DNA are more prone to mutations, and certain types of mutations are more likely than others. The fact that mutations are not entirely random, however, does not mean that some agent, biological or otherwise, directs at what time and in whom a particular mutation will occur. The notion that factors in the environment specifically direct the DNA to mutate in a particular way to allow organisms to adapt to the environment is contradicted by available evidence.

Some of this evidence comes from the fact that duplicated gene copies may mutate into useless pseudogenes, which remain as leftover by-products of gene duplication and divergence. Four of our thirteen copies of human globin genes are pseudogenes, three in the alpha cluster and one in the beta cluster.

We can tell approximately how old a pseudogene is by analyzing its sequence. Once a gene has mutated into a pseudogene, it no longer functions and is thus no longer subject to negative purifying selection. Any further mutations are neutral, so mutations tend to accumulate in pseudogenes as they age through the same process of mutational decay we encountered in the previous chapter. The more mutations a pseudogene has when compared to its parent gene, the older it must be.

Based on such an analysis, it is evident that the psi-beta pseudogene (shown in figure 6.2, page 162) is very old. This pseudogene has accumulated so many mutations relative to its parent gene that it must have initially mutated into a pseudogene more than one hundred million years ago. Further evidence of its ancient origin comes from the observation that all primates have it, or at least a piece of it, indicating that it mutated into a pseudogene in a common ancestor of all living primates.

By contrast, another globin pseudogene, the psi-zeta pseudogene in the alpha cluster, is one of two side-by-side copies of the zeta gene and is very young, almost identical to its next-door functional copy. It differs from the functional copy by a mere three mutations, one of which completely disables the gene. Chimpanzees also have two side-by-side copies of the zeta gene in the same place in their DNA as in ours, but both of their copies still function as bonafide genes. Sometime after the divergence of the lineages leading to humans and chimpanzees, a human copy decayed into a pseudogene while both chimpanzee copies remained functional.

These are not isolated examples. As it turns out, our genome is littered throughout with pseudogenes, many of them functionless mutated copies of parent genes that still function. In fact, we have at least as many pseudogenes as real genes.[13] About 30 percent of them arose when functional gene copies mutated into pseudogenes. The remaining 70 percent arose through a type of gene duplication that mimics the way viruses replicate, a fascinating story we'll explore in the upcoming chapter.

The evidence of gene evolution through duplication followed by divergence in our genome is abundant. Natural selection favored mutations in gene copies that confer a beneficial function to them, but not all copies mutated into something beneficial. Pseudogenes are mutated relics of this duplication-divergence process, and they offer evidence that mutations in gene copies are undirected. They persist because they confer neither a disadvantage nor an advantage, and in most cases a functional gene copy remains, often right next to the pseudogene.

In a few cases, however, even a pseudogene may be favored through natural selection, such as the human *CMAH* pseudogene, which allowed ancient humans to acquire resistance to malaria. Although this resistance lasted for about two million years, it turned out to be temporary because the parasites that infect us also evolve.

NEW GENES MAY EVOLVE THROUGH DNA REARRANGEMENT.

Another evolutionary process that creates new genes is DNA rearrangement. In the previous chapter, we saw how DNA in chromosomes may be shuffled through fusions, fissions, inversions, and other types of chromosome rearrangement. In rare cases, the breakpoints of a chromosome rearrangement may be within a gene, and the resulting DNA sequence at the breakpoint may bring pieces of different genes together to create what is known as a *fusion gene*, a gene consisting of segments from two or more preexisting genes. More common in gene evolution, however, are small-scale rearrangements that insert a portion of one gene into another, or that rearrange pieces of two or more genes to form a new fusion gene. In some cases, a piece of a pseudogene may be recruited and spliced to a preexisting gene to create a new fusion gene.

Not all fusion genes are beneficial, however. As you grow older, DNA rearrangements gradually accumulate in your cells. Each DNA rearrangement is a rare event, but given the fact that your body consists of trillions of cells, even very rare events are likely to happen in some of your cells as you age. Most of these rearrangements are benign, having no consequence whatsoever. They are evolutionary dead ends, dying out when you die. However, a few of them can be dangerous. For example, at least 95 percent of cases of a cancer known as chronic myelogenous leukemia result from a chromosome translocation in which pieces of chromosomes 9 and 22 exchange places. The resulting translocation fuses two genes, BCR and ABL, which are normally separated on different chromosomes. The BCR-ABL fusion gene is activated and overexpressed when it should not be; the end result is excessive cell division and cancer.[14] There are many other examples of cancer caused by fusion genes that arose through DNA rearrangements.

Not all fusion genes cause problems. Several of the normal functioning genes present in the genome of every human evolved as fusion genes. For example, scientists at the University of Bremen in Germany identified a fusion gene in humans that evolved when a functioning gene acquired part of a pseudogene to become a new fusion gene. They found this fusion gene in chimpanzee, gorilla, and lar gibbon genomes, indicating that it evolved at least thirty-five million years ago. Describing their discovery, these scientists wrote, "Usually considered a type of evolutionary waste, they [pseudogenes] melt into the background of their surrounding DNA by the loss of similarity to the active gene. . . . On the other hand, in this paper we describe the evolutionary recycling of this genomic waste."[15]

ENTIRELY NEW GENES MAY EVOLVE FROM DNA CONTAINING NO PREEXISTING GENES.

Although many of our genes have arisen from preexisting genes, a very few of our genes are entirely new and uniquely human genes, having arisen from favorable mutations in noncoding DNA since the divergence of the lineages leading to humans and chimpanzees. Scientists at the University of Dublin in Ireland identified three of these new genes in 2009 and predicted that a few other entirely new genes might be found (they estimated that the human genome might contain as many as eighteen). All three of these new genes are relatively short and simple, as is expected with newly arisen genes. The scientists compared these new genes with their corresponding sequences in other primates, including those of chimpanzees, gorillas, orangutans, gibbons, and rhesus macaques, and found that these sequences in other primates could not possibly function as genes. All the evidence indicates that these genes evolved recently through new mutations, and that they are unique to humans.[16] This discovery offers evidence that new genes may arise from noncoding DNA and that gene evolution is not entirely dependent on remodeling preexisting genes.

GENOME-WIDE COMPARISONS REVEAL HOW EVOLUTION TRANSPIRED.

Genome sequencing is a daunting task, although advances in technology are making it increasingly less so. Although it would be ideal to sequence the DNA molecule in each chromosome from one end to the other, DNA molecules from entire chromosomes cannot be kept intact for sequencing, nor can DNA-sequencing equipment sequence large molecules of DNA. Instead, scientists collect a large sample of random DNA fragments from the cells of an organism, a collection so large that the fragments span the entire genome, typically several-fold. They then sequence the fragments in high-speed automated DNA sequencers. Once the sequences are available, they use computers to assemble the sequences into their correct order on each of the chromosomes. Finally, they annotate the assembled genome, which means they identify and catalog the genes and other known functional and nonfunctional elements in the genome. So far, several primate genomes have been sequenced, assembled, and annotated—among them those of humans, common chimpanzees, the two orangutan species, and rhesus macaques. And several more are on the way.

Before genome sequencing, we could compare in detail only selected parts of the genomes among related species—any genome-wide comparison necessarily consisted of estimates based on extrapolation from samples. Now we can conduct true genome-wide comparisons in excruciatingly fine detail for those species whose assembled and annotated genomes are available. And the flood of information from these comparisons has told us much about how we evolved and how we are evolving.

The human and chimpanzee genomes are highly similar, the very result we expect when the genomes of two closely related species are compared. For the most part, they have the same genes in the same places in the same chromosomes and have a highly similar DNA sequence throughout, with an overall similarity of more than 98 percent, albeit with the ten large rearrangements we discussed in the previous chapter.[17] The most telling information, however, comes not from the similarities but from the differences. The human and chimpanzee genomes began diverging and accumulating different mutations when the two ancestral lineages split about five million to seven million years ago, which is merely a moment in evolutionary time. As we saw earlier, there are essentially three types of mutations—those that are beneficial, those that are harmful, and those that are neutral. Natural selection tends to preserve the beneficial mutations through positive selection and eliminate the harmful ones through negative purifying selection, while having little effect on neutral mutations. Hence, we expect to observe a gradual background accumulation rate of divergence for neutral mutations, a reduced rate for harmful mutations, and an accelerated rate for beneficial mutations.

Comparison of the human, chimpanzee, and rhesus macaque genomes has confirmed that the vast majority of genes are highly subject to negative purifying selection. In the human and chimpanzee genomes, 29 percent of all genes encode identical proteins, and most of the remaining genes encode proteins that are more than 99 percent similar, strong evidence of negative purifying selection.[18] When the rhesus macaque genome is added, the number of genes encoding identical proteins in all three species drops to 9.8 percent (which is still quite high), and the consistently lower number of mutations within genes compared to mutations overall confirms that negative purifying selection has been under way for most genes in all three species.[19]

Some of the most interesting genes are the outliers—those that have an unusually high number of non-neutral mutations when species are compared, which is the signature of positive selection. Genome comparisons among humans, chimpanzees, and rhesus macaques have revealed that certain genes are highly subject to positive selection in these three species. Not surprisingly, many of them are genes that confer resistance to infectious disease, such as malaria and tuberculosis; genes that govern reproduction; and genes related to the senses of smell and taste. A significant number of these genes are located within clusters of related genes that arose through duplication followed by divergence.[20] One of the traits that is most developed in humans is the size and function of our brains. Not surprisingly, genes that function in brain development have displayed a high rate of positive selection in the human ancestral lineage when compared to other primates and other mammals.[21]

Interestingly, much of the DNA bearing evidence of positive selection lies *outside* of genes. Throughout our DNA are regions that regulate genes, determining when they are turned on and off and to what degree their proteins are produced. Although our genes are highly similar to those in chimpanzees, many of the differences are alterations in gene regulation rather than in the genes themselves.

Comparison of our genome with genomes as close to ours as the chimpanzee and rhesus macaque genomes and as distant as the mouse, rat, and even zebrafish genomes shows that many of our genes are very old, present in all these distantly related vertebrate species. These genes continue to perform essentially the same function in all these species (including us) that they have performed for hundreds of millions of years.

Large-scale comparison of genomes confirms the mechanisms for gene evolution we just discussed. Some genes have simply evolved through the gradual accumulation of mutations. A large number of our genes evolved from other preexisting genes. Some arose from duplication followed by divergence, others when segments of preexisting genes were shuffled to form new combinations that became fusion genes. And an extremely small number of our genes evolved as new and uniquely human genes. Many of our older genes arose long ago as simple genes and have since evolved to become very large and complex.

The evidence from comparison of genes across genomes offers clear and overwhelming evidence of our shared evolutionary ancestry with other species, especially our primate relatives. The fact that we share most of our genes with other primates, that the physical arrangement of our genes in our chromosomes resembles the arrangements of genes in our primate relatives, and that we can trace the evolutionary origins of genes that differ among related species, can all be explained only through common ancestry followed by evolutionary divergence of species.

Although genes are the most important part of our genome, they account for a mere 1.8 percent of it. What about the remaining 98.2 percent of our DNA, the so-called noncoding portion? As it turns out, both positive and negative purifying selection have affected large portions of our noncoding DNA, in many cases within the regions that regulate how, when, and to what extent genes are turned on and off. Not only do mutations in our genes give us our uniquely human characteristics, mutations in the noncoding regions that regulate genes also are responsible for making us human. Even a gene that is identical in humans and chimpanzees may be expressed differently in the two species. In 2007, scientists at the Wellcome Trust Sanger Institute in England and the University of California–Santa Cruz and Pennsylvania State University in the United States identified more than 1,300 regions of DNA outside genes that bear the signatures of positive selection in humans, evidence that differences in gene regulation represent a major force in our evolution.[22]

Although these regions of gene-regulating DNA are numerous, they still represent just a small fraction of our genome. How did the rest of our genome evolve? In the next chapter, we explore the millions of wildly successful replicators in our DNA, invaders and freeloaders that are littered throughout our genome. They are unusually successful in primates, especially humans, accounting for nearly half of our DNA.

CHAPTER 7
EVEN MORE EVIDENCE FROM OUR GENOME

INVADERS AND FREELOADERS
BY THE MILLIONS

V iruses are at once frightening and fascinating. Among the most miniscule of biological entities, they straddle the boundary between the living and nonliving. They are so tiny that their images appear fuzzy at the highest magnifications possible with light microscopes—only powerful electron microscopes allow us to visualize them in detail. Outside of a cell, they act like nonliving crystalline structures, nothing more than a DNA or RNA molecule surrounded by protein and sometimes a tiny bit of fat. In this state, a virus is silent and inert, with no metabolism, reproduction, or growth. It is incapable of expending any energy, which prevents it from moving on its own. Instead it must depend on wind, water, body fluids, or other agents to carry it.

The overwhelming majority of the countless virus particles everywhere around us remain inert, never causing infection. Their destiny is silent and uneventful degradation. We can't even say that they die because, outside of a compatible host, they're not really alive. A small fraction, however, happen to make their way into living organisms, as when someone breathes air or ingests liquids or food containing viruses. When one of those lucky virus particles makes contact with the cells of the unlucky organism that happens to be a suitable host, the virus suddenly springs to life. It attaches itself to the surface of the cell and enters the cell. Eventually the cell may unwittingly make many copies of the virus, and these newly replicated virus particles escape the cell to attack other cells, propagating the infection.

Some of the most devastating illnesses in human history—such as smallpox, polio, influenza, and AIDS—are viral diseases. Moreover, many viral infections are either difficult to treat or cannot be treated. We can prevent some of them with immunizations, but in most cases, once the infection has taken hold, we must let the disease run its course until the body's immune system defeats the virus or the infected individual dies. Fortunately, treatments for certain viral diseases are now available. But rarely do the treatments eliminate the viruses— they merely treat the symptoms or hold the viruses at bay.

Beyond the health threat viruses pose for all of us, they have played a major role in the evolution of our genome for tens of millions of years. And a few of them are still at it.

ANCIENT VIRUSES ARE TRAPPED IN OUR GENOME.

During the early days of genome sequencing, when scientists began sequencing small segments of the human genome, they found remnants of ancient, now extinct, viruses scattered throughout it. These viruses are called *human endogenous retroviruses*, or HERVs, for short. They are remnants of *retroviruses*— viruses that use RNA instead of DNA as their genetic material. The *retro* in the word *retrovirus* implies that something is reversed. *Transcription* is a routine and essential process for our cells through which they make RNA copies of genes from the DNA of those genes, the first step in the DNA-to-RNA-to-protein transfer of information. However, when a retrovirus infects a cell, transcription is reversed—the virus's RNA is copied backward into a molecule of DNA, and the DNA copy of the virus inserts itself into the cell's DNA. The DNA copy, now embedded in the cell's genome, serves as the master pattern for encoding new RNA copies of the virus.

The viral infection destroys many of the cells that contain a DNA copy of the virus. The body's immune system also destroys many of the infected cells by selectively attacking them. Those cells that manage to persist with viruses inside them typically die soon after the organism dies. The virus survives only by escaping cells it has infected and going on to infect other cells, often in other individuals.

However, millions of years ago in our distant ancestry, certain retroviruses managed to infect the *germline cells*, which, as we discussed in chapter 5, are cells in the reproductive organs that eventually develop into egg or sperm cells. If a DNA copy of a virus happens to be integrated into the DNA of a germline cell, and an egg or sperm arises from that cell, that egg or sperm carries the virus embedded as part of its DNA. If that egg or sperm cell then participates in fertilization, the resulting individual inherits a DNA copy of that virus *in every cell of its body*. It will pass the virus to its offspring, who then pass it on to their offspring, and so on, generation after generation—not through infection but through inheritance.

The virus now has a new way to persist. Instead of relying solely on its ability to escape cells and infect other cells to propagate itself, the cells themselves now propagate the virus as part of normal DNA replication. If the virus has inserted itself at a place in the DNA where it causes no harm to the individual who carries it, it is selectively neutral. From that point forward, the virus may remain forever present in the genomes of potentially countless generations of individuals who descend from that original individual who was infected.

Genome sequencing has shown that several retroviruses infected the germlines of our remote ancestors at different times and have persisted in our genome to the present. Some of these remnants of ancient viruses are very old, having entered our ancestral genome tens of millions of years ago. A few are more recent, but the youngest are still at least a million years old. None have entered our genomes recently. In fact, all of the HERVs we now have in our genome no longer function as infectious viruses, having lost their infectivity long ago. Instead, they now exist only as relics of once-active viruses, trapped in our genome like molecular fossils. They invaded the cells of our remote ancestors as infectious viruses, but now they are so mutated that they can no longer do so. However, as we are about to see, they have yet another way of further invading our genome.

HERVS MAY RETROTRANSPOSE.

All humans have HERVs in our cells—lots of them—about 443,000, which is about twenty-two times the number of genes we have. They constitute 8.29 percent of our DNA.[1] This number is large because some HERVs can *retrotranspose*, which means that they make RNA copies of themselves, and then our cells recopy them as DNA. When a HERV retrotransposes, the original copy stays in place, and a new DNA copy is inserted elsewhere in the genome, increasing the number of copies of that HERV in the genome by one.

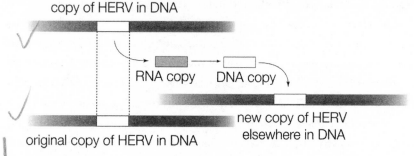

copy of HERV in DNA

RNA copy   DNA copy

original copy of HERV in DNA   new copy of HERV elsewhere in DNA

Even though each retrotransposition event adds only one additional copy to the genome, over long periods of evolutionary time, tens of thousands of retrotransposition events have added large numbers of HERVs to our genome, now inherited by everyone on the planet. The 443,000 copies of HERVs in our genome are the result of a few original infections followed by multiple rounds of retrotransposition occurring repeatedly over millions of years.

Fortunately, retrotransposition of HERVs is now a rare event. Most HERVs are so highly mutated that they can no longer retrotranspose—each remains stuck and silent where it resides, a relic of a time long gone when it inserted itself into the DNA of one of our remote ancestors. However, a very small number of HERVs are still active and can retrotranspose. When they do, they can cause inherited disorders and cancer, especially if they happen to insert themselves into or near a gene.

Because of the mutations they carry, all HERVs have been extinct as infectious retroviruses for a very, very long time. Is there any chance that a HERV could mutate back into a functional virus—a resurrection, so to speak, from its genomic grave in our DNA? Two groups of researchers, one in France and the other in the United States, decided to find out.[2] Both groups compared the many copies of some of the youngest HERVs with one another and came up with what we call a *consensus sequence*—the sequence of DNA that is most representative when many mutated copies of a particular DNA sequence are compared. If HERVs were once ancient viruses, the mutations they carry probably happened after they inserted themselves into the genome, and a mutation in any one HERV copy was probably independent of a mutation in any other copy. Therefore, the consensus sequence probably is—or at least comes very close to—the original sequence of the virus when it was infectious.

After identifying consensus sequences of these youngest HERVs, the teams of researchers artificially reconstructed segments of DNA with the consensus sequences and inserted them into cultured human cells. The cells produced copies of resurrected viruses that were capable of infecting other cultured cells. The French team named the virus they produced *Phoenix*, after the mythical bird that, after its death, arose from the ashes to live again. They also found that it is possible, through recombination of HERV sequences already in our genome, for a virus like Phoenix to arise naturally without any scientific intervention, although there is no evidence that one has ever done so.

At first glance, this sort of research seems frightening—almost like something from a science fiction horror movie. When the story of these resurrected viruses broke in 2007, some questioned the wisdom of reconstituting an ancient virus because of the danger it might pose to human health should it escape the laboratory. However, modern humans are genetically different from our distant ancestors who first acquired these viruses. Because of these differences, the chance of this virus successfully causing a serious disease in modern humans was only a remote possibility but still of sufficient concern for the scientists to conduct their research under strict conditions of biosafety and isolation—the same conditions as those applied when scientists research modern human viruses. These reconstituted viruses infect only cultured cells under very strict conditions that are possible only in a laboratory; they cannot infect modern humans, although their ancient ancestors infected our ancient ancestors. Even so, like all human viruses cultured in laboratories, they are carefully guarded.

Far more dangerous are modern viruses that infect millions of people worldwide, such as influenza and HIV. Researchers routinely study these viruses under strict biosafety and isolation to help us find better treatments, vaccines, and ways to prevent the viruses from spreading.

RETROELEMENTS CONSTITUTE THE LARGEST FRACTION OF OUR GENOME.

Human endogenous retroviruses are not the only segments of DNA capable of retrotransposition. They are members of a large group of sequences in our genome called *retroelements*—sequences of DNA that retrotranspose, just as HERVs do, to replicate themselves in our genome. We can think of HERVs as ancient invaders that entered our genome long ago; a few of them still invade new parts of our genome as they retrotranspose. There is, however, another group of retroelements that function as highly prolific replicators— pieces of DNA that have just the right sequences to proliferate like HERVs through retrotransposition. They act like HERVs but without all the extra viral baggage that HERVs have.

Some of them contain just two genes, which are needed to carry out their own retrotransposition. These elements are called LINEs, for long interspersed nuclear elements—*long* because they are relatively long compared to most ret-roelements, and *interspersed* because they are scattered all through our genome. We have about 868,000 copies of LINEs in our genome, constituting more than 20 percent of our DNA.[3]

Another class of retroelements are SINEs. As you might guess, SINE stands for short interspersed nuclear element. They are the ultimate free-loaders, possessing none of the genes needed to retrotranspose. Instead, they piggyback on the products made by LINEs and HERVs to carry out their retrotransposition. They are wildly successful replicators, as evidenced by the sheer number of them in our genome—more than 1.5 million copies—consti-tuting a little more than 13 percent of our genome.[4]

Yet another set of retroelements is found in our genomes, and, in a strange way, these retroelements are closely related to our genes. Recall that RNAs are transcribed from genes and that the RNAs are then translated into pro-teins. Occasionally, the same proteins that retrotranspose HERVs, LINEs, and

SINEs hop onto an RNA made from a gene and retrotranspose the gene's RNA into DNA, then insert the DNA copy into the genome as a new pseudogene.

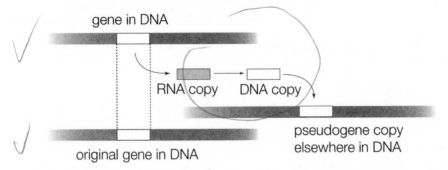

We know this because RNAs are somewhat different than the genes that encode them. Most of our genes in DNA have what are called exons and introns. The *exons* are the essential parts of the genes, and they end up in the final RNA. The *introns*, by contrast, are segments of DNA that are present in genes but are cut out of the RNAs after they are transcribed from the gene as the exons are spliced together. For example, let's examine the *metallothionein-2A (MT2A)* gene, which protects our cells from heavy-metal toxicity. With a mere 932 base pairs, the *MT2A* gene is quite small and is thus relatively easy for us to visualize—most human genes are large and complex, with tens to hundreds of thousands of base pairs in them.

In figure 7.1 (page 180), (a) shows a diagram of the DNA in the human *MT2A* gene and (b) shows the RNA transcribed from this gene. Like the vast majority of human genes, this gene contains introns, which are cut out of the RNA after it is made, so the RNA is shorter than the DNA that encoded it. Remember that RNA, like DNA, consists of a linear sequence of bases, U, C, A, and G in RNA. As the RNA diagram illustrates, our cells tack a long string of As onto the RNAs they make from genes, a feature called a *poly(A) tail*. The poly(A) tail is not encoded in the gene's DNA but is added after the RNA is made.

We can spot retroelements made from genes because the introns are missing, and because they often have a poly(A) tail in their DNA, which real genes do not have. These gene-like retroelements are often called *retro-pseudogenes*—*retro* because they are made by retrotransposition, and *pseudogenes* because nearly all of them look like their parent gene, although most of them fail to function as genes.

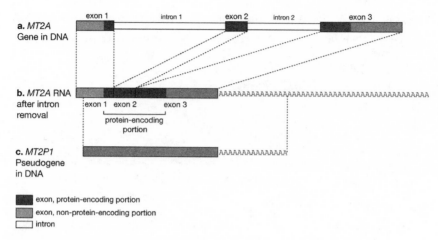

Figure 7.1. Comparison of the *MT2A* gene in humans with the RNA it encodes and a retropseudogene derived from it. *(a)* The DNA of the gene contains three exons and two introns. *(b)* The introns are removed, the exons are spliced together, and a poly(A) tail is added to form the final RNA. *(c)* The retropseudogene, *MT2P1*, resembles the RNA in that it has no introns and has a poly(A) tail.

Figure 7.1 (c), above, shows the *MT2P1* retropseudogene, which is derived from the *MT2A* gene's RNA. The entire DNA sequence of this gene, its RNA, and its pseudogene are small enough for us to compare them in complete detail and fit them on a single page in a book, which is what figure 7.2 does. Notice that in figures 7.1 and 7.2 that the *MT2P1* retropseudogene is very similar to the *MT2A* RNA, with a poly(A) tail and no introns. However, segments are missing from both ends of the pseudogene, and seventeen mutations have accumulated in it since its insertion (outlined by boxes in figure 7.2). Overall, this retropseudogene differs by 4 percent from its parent gene.

Although not as abundant as LINEs and SINEs, retropseudogenes are nonetheless quite prevalent in our genome, numbering more than 14,000.[5] They are especially abundant in mammals but are extremely rare or absent from the genomes of most nonmammalian species. As an example, let's look at the genes that regulate glycolysis, an essential energy conversion process that most cells in our body carry out. Some of the most abundant and extensive analyses of retropseudogene evolution have come from Mark Gerstein's

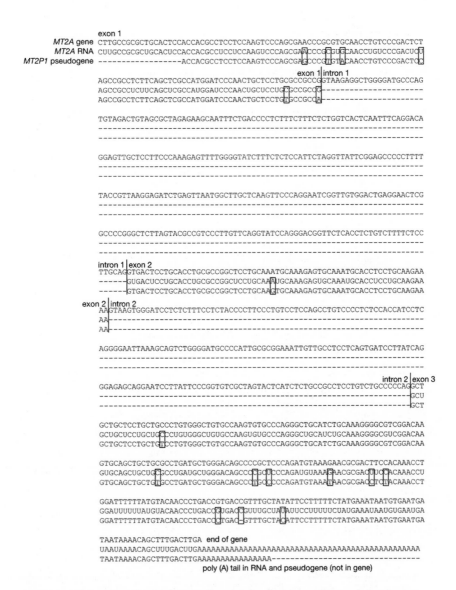

Figure 7.2. Aligned sequences of the *MT2A* gene, its RNA, and the *MT2P1* pseudogene. The boxes represent mutations in the pseudogene when compared to the parent gene and its RNA. In the RNA, the base U is present wherever there is a T in the DNA. Although derived from the RNA, the pseudogene contains Ts instead of Us because it consists exclusively of DNA.

laboratory at Yale University. In 2008, scientists working at this laboratory published a survey of retropseudogenes governing glycolysis in the human, chimpanzee, mouse, rat, chicken, zebrafish, puffer fish, fruit fly, and nematode worm genomes. Retropseudogenes generated by these genes were abundant in the human, chimpanzee, mouse, and rat genomes but were entirely absent from the others, a strong indication that retropseudogenes, as well as most other retroelements, are by far most prevalent in mammalian genomes, including the human genome, and are highly important in mammalian evolution.[6]

Evolutionary theory predicts that for a retropseudogene to be established in the genome and inherited, it must be inserted in the germline and then end up in an egg or sperm cell that undergoes fertilization. Any retropseudogene inserted in a cell that is *not* in the germline (which includes the overwhelming majority of our cells) does not persist beyond the life of the individual. Although evolutionary theory predicts that retropseudogenes should consist of genes turned on in the germline, theories are strongest when supported by reliable experimental evidence. In 2003, scientists from Gerstein's laboratory published a compilation of human retropseudogenes identified when they surveyed the entire human genome. Not surprisingly, they found that retropseudogenes are indeed derived from germline-expressed genes, many of them from so-called housekeeping genes, genes that are essential for basic functions in nearly all types of cells, including those in the germline.[7] The genes governing glycolysis, as we just discussed, are examples of housekeeping genes.

HOW IMPORTANT ARE RETROELEMENTS?

Altogether, the retroelements—HERVs, LINEs, SINEs, retropseudogenes, and a few others—are the largest class by far of any group of DNA elements in our genome, comprising almost 44 percent of our total DNA. Most of them serve no known function, but there are exceptions. A few have been co-opted by the genome to assist in regulating genes. Some have inserted themselves into genes and have inactivated them. A few retropseudogenes have developed into functional genes called retrogenes, some of which are implicated in cancer.[8]

And a very few retroelements have fortuitously reassembled themselves into sequences that function as entirely new genes in our genome.[9] However, as far as we can tell, the vast majority of retroelements simply reside where they happened to insert themselves, benignly replicating every time the DNA in our cells replicates.

So how important are they? In terms of sheer quantity, they are extremely important. However, in terms of function, they appear to be much less important than other parts of the genome. Most of them appear to be invaders and freeloaders, doing little to help us and sometimes a bit to hurt us (by causing some cases of genetic disorders and cancer), but mostly they are just freely riding along in our DNA, being replicated every time our DNA replicates. There is good evidence that most of them do essentially nothing, neither helping nor harming us, persisting generation upon generation for millions of years.

WHAT DO RETROELEMENTS TELL US ABOUT OUR EVOLUTION?

Because they are so abundant, retroelements tell us an enormous amount about how they, and our genome, evolved. First, the vast majority of retroelements inserted themselves into our genome millions of years ago, then decayed into inactivity as mutations accumulated in them. Most are now completely inactive, incapable of retrotransposing ever again. We can determine how old these retroelements are by comparing them with the young ones—the ones that are still retrotransposing. Because mutations accumulate in a retroelement over time, the number of mutations one carries is a good indication of how old it is. A highly mutated retroelement must be very old, whereas one with just a few mutations must be relatively young. Without reference to any other species, scientists can analyze the DNA of retroelements in the human genome and determine how old they are.

Then, if they extend this analysis to other related species, those retroelements identified as the oldest should be present in the same places in the genomes of related species because they were inserted into the genome of a distant common ancestor of those species long ago. Younger elements, however, should be present only in closely related species because they inserted themselves in a more recent common ancestor. The youngest elements should be present in just a single species.

For the most part, the retroelements in our genome originated since the emergence of primates, so they are especially valuable for studying primate evolution. When the retroelements in our genome are compared with those of our closest primate relatives, such as chimpanzees, gorillas, and orangutans; more distant primate relatives, such as monkeys, lemurs, and galagos; and to a lesser extent with other mammals, such as mice and rats, all these predictions are met—hundreds of times over, offering spectacularly abundant evidence of our evolution.

Let's look first at how a SINE named *Alu* evolved. It is by far the most abundant retroelement in humans and in our primate cousins, with more than a million copies making up a little over 10 percent of our genome.[10] Because *Alu* elements are found only in primates—and all primates have them—they probably originated about ninety million years ago in a primate common ancestor. By comparing the oldest-known *Alu* elements with progressively younger ones, including the youngest, which are still retrotransposing, scientists have reconstructed the story of how they evolved.

Certain genes are very small and encode only RNA—they are too small to encode proteins. A group of them encodes a very small RNA with less than a hundred base pairs called 7SL. About ninety million years ago, one of these genes became the first *Alu* element in an ancient ancestral primate. According to fossils of early mammals alive at the time, it probably was a tree-dwelling rodent-like species about the size of a squirrel, living at the time when dinosaurs flourished. This earliest *Alu* element was very small, little more than a hundred base pairs of DNA. It piggybacked on LINEs to retrotranspose— LINEs are much older than *Alu*, as evidenced by their presence in both primate and rodent genomes. A few of these oldest *Alu* elements are still identifiable in our genome, relics of that time long ago when dinosaurs still roamed the earth.

This first *Alu* element apparently was not an effective replicator, but it still managed to proliferate very slowly. Eventually, copies of it diverged into two groups, one called FLAM, for free left *Alu* monomer, and the other called FRAM, for free right *Alu* monomer. Then a major event in *Alu* evolution happened. A FLAM fused with a FRAM, and the resulting FLAM-FRAM fusion element evolved into an enormously successful replicator, the distant ancestor of *Alu* elements.[11]

By then, the earliest primate lineages were just starting to evolve. In our most distant primate relatives, the prosimians (lemur, galago, marmoset, and others), *Alu* elements have steadily increased in number at a relatively constant rate. However, in our own ancestry, there was an enormous burst of *Alu* retrotransposition, producing hundreds of thousands of copies, about thirty-five million years ago. This burst predated the divergence of Old World monkeys from humans and apes, so we share an enormous number of *Alu* elements with these species—hundreds of thousands of them—in exactly the same places in our genome and theirs.[12]

Then, if we fast-forward to just a couple of million years ago, after the human and chimpanzee ancestral lineages had diverged, there was another burst of *Alu* activity in the human ancestral lineage, albeit more modest than the earlier burst. As a consequence, more than 99 percent of the *Alu* elements in our genome are also present in the chimpanzee genome in the same places, in the same orientations, and with nearly all the same mutations when compared to their original parent elements. But we have about twice as many newly inserted *Alu* elements as chimpanzees because of this modest retrotransposition burst that happened exclusively in our ancestry.[13] This more recent burst has since dissipated. *Alu* element insertion still happens in the human genome—freshly inserted *Alu* elements are discovered in people quite often—but at a considerably slower pace than in our evolutionary history.

Genome-wide studies on retroelements, like those we've just reviewed, are especially informative because they show us how our genome has evolved on a broad scale. However, because many of these studies examine thousands to millions of retroelements, it is not easy to visualize how it all happens. The scientific literature is replete with elaborate, complex, and highly informative studies, analyzed with computer software specifically written for retroelement analysis on a large scale. These studies invariably reveal a coherent and predictable evolutionary pattern. To see that pattern, let's turn our attention not to a large-scale, complex example but to one that is straightforward, simple, and easy to visualize.

Earlier, we looked at the *MT2A* gene and its retropseudogene, *MT2P1*, as an example of how retropseudogenes originate. The information in figure 7.2 (page 181), tells us that this retropseudogene differs from its parent gene by 4 percent due to mutations that have accumulated since its insertion. According to evolutionary theory, mutations in pseudogenes should typically be selectively neutral—natural selection should have no effect on them—because most pseudogenes are nonfunctional. Genome wide, the human and chimpanzee genomes differ in selectively neutral regions, which constitute most of the two genomes, on average by about 2 percent. The fact that the *MT2P1* differs by twice this much suggests that its origin probably predates the divergence of the ancestral lineages leading to humans and chimpanzees. The number of mutations it carries suggests that it's at least ten million years old. If this reasoning is correct, we ought to find this same pseudogene in both the human and chimpanzee genomes at exactly the same place. While writing this chapter, I took a break and searched the chimpanzee genome for the *MT2P1* pseudogene. Within a few seconds of searching, I found it exactly where I expected it to be, on chimpanzee chromosome 4 at a site corresponding precisely to the place where it resides in the human genome.

Now, let's take this reasoning a step further. The rhesus macaque genome differs in selectively neutral regions from the human genome by about 8 percent on average. The *MT2P1* retropseudogene differs from its parent gene by half this amount, so we expect it to be younger than the time when the human–great ape and Old World monkey ancestral lineages diverged, about twenty-nine million years ago. By this logic, *MT2P1* should be missing from the rhesus macaque genome. When I searched the rhesus macaque genome, I found the corresponding site where the *MT2P1* pseudogene is present in the human and chimpanzee genomes, and, as predicted, *MT2P1* is absent from the rhesus macaque genome.

This is not an isolated example. I chose it because of its simplicity and historical importance (*MT2P1* was one of the first retropseudogenes discovered). I could have chosen any number of other gene-retropseudogene pairs and determined approximately when the retropseudogene arose on the basis of mutations, and the predictions of which species have it and which do not would have been met.

Certain retropseudogenes offer us a unique opportunity to study evolution because a single parent gene has generated a relatively large number of pseudogenes, which were inserted into a genome over long periods of evolutionary time. Let's focus on an example I used in my book *Relics of Eden*—the *NANOG* gene, with its pseudogenes and *Alu* elements. Since publication of that book, the story has unfolded in a fascinating way, which I'll update here.

The *NANOG* gene itself, independent of its pseudogenes and *Alu* elements, is quite intriguing and is currently one of the hottest genes for scientific research. To understand it, we need to go to the earliest stages of our lives. Most of us have at one time or another heard the expression "Think back to when you were a child." For fun, I occasionally ask students in my biology classes to "think back to the time when you were a single cell." Of course none of us can do so because at the time, we had no brains and no nervous system, and we were completely incapable of thinking. When an egg cell from your mother united with a sperm cell from your father, you were, for a very brief time, a single cell. That cell divided to form two cells, which divided to form four, and so on, until you were a ball-shaped structure of genetically identical cells. Within this ball was a mass of cells that were destined to eventually become your body, with all the anatomical features that now define you. Those cells were embryonic stem cells.

Most of us have heard of embryonic stem cells because of the political debate surrounding them. We'll not get into that debate here but will focus instead on the biological role of these cells. Embryonic stem cells are what we call *undifferentiated cells*, which means that they have not yet started to develop into any particular body part. Instead, each one is capable of developing into any body part under the right conditions, which is why they are so important for medicine. Potentially, embryonic stem cells can repair damaged parts of the body by regenerating their tissues, just as they are capable of generating them for the first time.

It is at these earliest stages—when embryonic stem cells have not yet started to differentiate—that the *NANOG* gene plays its role. It is one of several genes that prevent embryonic stem cells from differentiating. When *NANOG* and these other genes are turned on, they maintain embryonic stem cells in an undifferentiated state until enough of them have proliferated to successfully begin differentiating into an embryo. In a sense, they keep embryonic stem cells in their "youthful," undifferentiated state, which is how the gene got its name. *NANOG* is derived from the Celtic word *Tir-Nan-Og*, which is the name of the ancient land of eternal youth in Celtic mythology— and a popular name for Irish pubs.

On a wet spring day, I was at home, keeping a close eye on my autistic son. I had my laptop and was perusing scientific articles on the Internet— not the usual activity for most people on a Saturday afternoon, but it's one I happen to immensely enjoy. I came across an intriguing article titled "Eleven Daughters of *NANOG*" by Anne Booth and Peter Holland, scientists at Oxford University.[14] In it, they documented eleven pseudogene copies of the *NANOG* gene in the human genome. All these pseudogenes had arisen independently from the *NANOG* parent gene. Booth and Holland determined the evolutionary ages of these pseudogenes, from oldest to youngest, by comparing them to the parent gene under the assumption that the more mutations a pseudogene has accumulated when compared to its parent gene, the older it must be.

At that time, the chimpanzee genome had just been released on the Internet but had not yet been published. I began searching through the chimpanzee genome to see what I could find about these pseudogenes. I found ten of them in exactly the same places in the chimpanzee genome as in the human genome, just as expected if they had been inserted into the genome before the divergence of the ancestral lineages leading to humans and chimpanzees.

I couldn't find one of them—the one named *NANOGP8*—in the chimpanzee genome. To be sure it was truly missing, I checked to see if there might be a gap in the chimpanzee genome assembly, a possibility in early genome assemblies when small portions of the overall sequence have yet to be determined. I looked in the chimpanzee genome for the site where *NANOGP8* had been inserted in the human genome and had no trouble finding it. There was no gap there—*NANOGP8* was clearly missing from the chimpanzee genome.

The reason was immediately obvious. This pseudogene differs from its parent gene by less than 1 percent, which means it is a very young pseudogene. Booth and Holland had determined that *NANOGP8* was the youngest of the human *NANOG* pseudogenes. All lines of evidence prior to my search pointed to the conclusion that *NANOGP8* should not be in the chimpanzee genome—which is exactly what I found.

The story then became more intriguing. It turns out that the functional *NANOG* gene, which is the parent of the pseudogenes, has several *Alu* elements, or pieces of them, embedded in it. All but one of these *Alu* elements are in the introns, and the one that's not is in a part of the gene that does not encode the protein. So all these *Alu* elements have no effect on the gene's function. They are stuck there as extra baggage, tagging along inside the gene every time it gets replicated. Booth and Holland noticed that the single *Alu* element not in an intron is missing from all the *NANOG* pseudogenes except *NANOGP8*, which happened to have that particular *Alu* element in it. They concluded that this *Alu* element inserted itself into the *NANOG* gene before *NANOGP8* was made but after all the other pseudogenes had been made.

Did this *Alu* insertion precede the divergence of the human and chimpanzee ancestral lineages? I looked at the chimpanzee *NANOG* gene, and the *Alu* element is there. It must have been inserted into the *NANOG* gene before the divergence of the ancestral lineages leading to humans and chimpanzees but after insertion of all *NANOG* pseudogenes, except *NANOGP8*.

After considerably more research, a colleague and I published a detailed evolutionary history of the human and chimpanzee *NANOG* gene and its pseudogenes in 2006.[15] The following year, the rhesus macaque genome was released, and once again I found myself searching for *NANOG* pseudogenes. As with the chimpanzee genome, I found ten *NANOG* pseudogenes in the rhesus macaque genome. However, only seven of them were present in the same places as in the human and chimpanzee genomes. Evolutionary theory predicts that when some but not all pseudogenes are shared between related species, the shared pseudogenes should be the oldest, and that's what I found. The remaining three were young, inserted into the macaque genome after the ancestral lineage leading to the macaque diverged from the one leading to humans and chimpanzees. Figure 7.3 (page 190) summarizes the evolutionary history of the *NANOG* pseudogenes in these three lineages.

Recently, scientists have discovered that the human-specific *NANOG* pseudogene, *NANOGP8*, is apparently detrimental to humans. In most cells, it is completely silent. But it is active as a retrogene in some cancer cells,

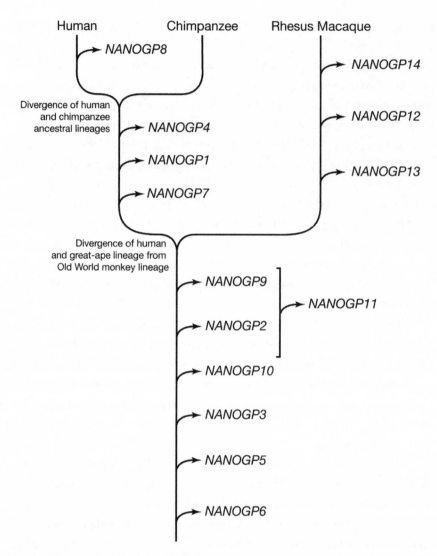

Figure 7.3. Evolution of the *NANOG* pseudogenes in the human, chimpanzee, and rhesus macaque genomes. The seven oldest pseudogenes are shared by all three species and are in exactly the same places in the three genomes. They originated before the divergence of the human and great ape ancestral lineage from the Old World monkey ancestral lineage.

encoding an RNA and a protein nearly identical to the *NANOG* protein, and it promotes the development of cancer.[16]

This personal research story I've just recounted nicely illustrates the process of discovery that scientists experience when studying genome evolution. In a way, it's like a mystery, filled with clues that allow us to solve an evolutionary puzzle. Those of us who do it find the fascination of discovery addictive. And it is research we can do anywhere we have a computer with an Internet connection—no laboratory required. However, the story I've just told does not recount a major discovery; rather, it is one of hundreds that conform to the same general evolutionary history.

Our genome offers profuse evidence that we evolved along with the rest of life and that the same processes that shaped our genome in the past are still under way. Evidence that we're still evolving comes from the discovery that HERVs, LINEs, SINEs, and pseudogenes are arising anew in people today, just as they have for tens of millions of years.

But does any of this really matter? With disease, hunger, poverty, energy depletion, environmental degradation, and so many other problems facing humanity, shouldn't biologists be devoting their research to solving them rather than studying evolution? This argument is a common one, often touted by opponents of evolution who use it to attack evolutionary science from a different angle—instead of disputing evolution itself, they argue that teaching it is a waste of time.

Basic science is absolutely essential; it does not need to be applied to be of value. It consists of research aimed purely at discovering the wonders of nature from the most miniscule subatomic particles to the vast reaches of the intergalactic universe, including life on the earth. Applications of discoveries in basic science are not what drive the research but rather the intrinsic value of expanding our understanding of nature. Valuable applications, however, are often a benefit of basic science. In fact, all the applied sciences are founded in basic science. We can look at basic science as having two major roles. First, it provides us with a fundamental understanding of nature and how we are a part of it. Second, it serves as the foundation for applied scientific research, which is aimed at improving our lives.

The applied sciences are broad, including medical, agricultural, environmental, and technological research. They address our health, our food, the buildings that house us, the vehicles we use for transportation, the various technologies we utilize, and the environment in which we live. At first glance, they might seem to have little to do with evolution. However, basic research in evolution is absolutely essential to many of the applied sciences, particularly to medical, agricultural, and environmental science.

Throughout my career, I've had one foot in basic science and the other in applied science, which is not unusual for scientists. Much of my applied research has focused on the genetics of food plants cultivated in various parts of the world, with an emphasis on helping impoverished people overcome hunger and malnutrition. Having applied evolutionary theory in my research, I can unhesitatingly say that evolution *never mattered more* than it does now, as we address pressing issues of health, hunger, poverty, and our environment. The remaining chapters of this book show why.

CHAPTER 8
EVOLUTION AND OUR HEALTH

To make an important point in my classes, I often ask my students to raise their hands if they have fears of contracting smallpox, diphtheria, typhoid fever, whooping cough, tuberculosis, plague, polio, or cholera. Only rarely does a hand go up. I then ask how many have fears of getting cancer, Alzheimer's disease, diabetes, or heart disease, and nearly every hand goes up. The first set of diseases are infectious—caused by bacteria, viruses, or other microorganisms. The second set, for the most part, consists of noninfectious diseases—those related more to age, genetic predisposition, abiotic (nonbiological) agents, and lifestyle.

Had those same questions been addressed a century ago in a class of American university students, the response would likely have been the reverse. Throughout most of human history, infectious disease has been one of the most common causes of death and suffering, and it still is in many parts of the world. Even in war, soldiers often were more likely to die of infectious disease than in battle.

Today, modern medicine has dramatically reduced the incidence of many infectious diseases. To be sure, some are still serious. Combating the global AIDS pandemic, caused by the human immunodeficiency virus (HIV), has been a major challenge to scientists and healthcare professionals. And some of the diseases mentioned in the first paragraph still afflict millions of people every year worldwide. For example, only about four hundred people contract typhoid fever each year in the United States, most after traveling abroad. Yet there are more than twenty-one million cases of typhoid fever globally, most due to unsanitary conditions.[1] Even so, thanks to clean drinking water, sanitary food handling, antibiotics, immunizations, routine medical care, and other health interventions, infectious disease has been considerably reduced, especially in countries with extensive public health programs and adequate water and sewage infrastructure.

EVOLUTION PLAYS A BACK-AND-FORTH GAME BETWEEN US AND THE MICROBES THAT INFECT US.

Near the heart of Paris is the Pasteur Museum, a small but excellent museum that, unfortunately, only a handful of the millions of tourists visiting Paris choose to explore. Next door is the Pasteur Institute, one of the most advanced biomedical research facilities in the world and the site of some of the most significant discoveries in biology and medicine. Both are named after Louis Pasteur (1822–1895), the famous French scientist who was one of the principal founders of the science of microbiology.

In the museum, which was Pasteur's home and workplace, much of his laboratory has been preserved with its original instruments and glassware. Among the items of glassware are swan-necked flasks with long, narrow spouts

curving downward but remaining open to the air (figure 8.1). Pasteur filled swan-necked flasks with beef broth, then he boiled the broth. Because the spout remained open to the air, no bacteria grew in the broth, even after weeks or months.

On the other hand, if he broke the swan neck, so that its opening faced upward, the broth rapidly became overgrown with bacteria. The reason? Boiling the broth killed bacteria that were present in it. However, bacteria are ubiquitously present in the air. When the top of the flask is broken and

Figure 8.1. A swan-necked flask similar to the one used by Pasteur in his experiments.

open to the air, bacteria fall from the air into the flask and begin growing in the nutrient-rich broth. When the swan-necked spout is intact, air can move through the spout, but the airborne bacteria become trapped in the curve so they cannot enter the flask.

Pasteur's experiment decimated the notion of *spontaneous generation*, the idea that life spontaneously and repeatedly originates from nonliving substances. In Pasteur's experiments, bacteria came only from previously living bacteria. At the time of Pasteur's experiments, belief in spontaneous generation was already in decline. But for much of history, people thought that mushrooms, mice, insects, and other organisms came to life spontaneously, formed anew and instantly from nonliving material. By the mid-nineteenth century, although many rejected spontaneous generation of plants and animals, they believed that bacteria did arise spontaneously from nonliving material because they appeared to do so when viewed under a microscope. Pasteur's experiments, along with others, dispelled this notion.

This was but one of several major contributions Pasteur made to biology and medicine. He conducted experiments confirming the germ theory of disease—that infectious disease is a consequence of microbes entering a person (or animal or plant) and multiplying, causing disease symptoms in the process. He developed a procedure called pasteurization, named in his honor, in which milk, fruit juice, or other liquid food products are heated to kill microorganisms that live in them to make the products safer for human consumption. Today, most dairy products and fruit juices are pasteurized.

Pasteur was also a pioneer in vaccine development, which is dependent on our immune system. Immune systems in animals began evolving several hundred million years ago through gene duplication followed by gene divergence. That system has continued to evolve into a complex and elaborate group of genes, some of which are capable of rearranging themselves to produce products that recognize an enormous range of chemical substances that are foreign to our bodies. Because pathogens (disease-causing microorganisms, such as infectious bacteria and viruses) are foreign to our bodies, they carry substances that certain cells in our immune system recognize as foreign. Those cells add a chemical label to the foreign substances, and in so doing, they label the invading pathogens. Other cells in the immune system recognize the labels on the pathogens and destroy them. Once our immune system has attacked and destroyed a particular pathogen, it retains the ability to quickly relabel and attack that same pathogen should it again enter the body, rendering us immune to further infections by that particular pathogen.

Vaccines are weakened bacteria or viruses that are no longer able to cause serious illness. Or, in some cases, they are closely related species of bacteria or viruses that cause disease in animals—but not in humans, though they still elicit an immune response that protects a vaccinated person against the human disease. For example, in Pasteur's day, it was well known that the cowpox virus, which does not cause human disease, serves as an effective vaccine against smallpox. It stimulates the immune system in humans to produce antibodies that label the harmless cowpox virus for destruction, and these same antibodies also label the related smallpox virus, thus immunizing people against smallpox. This cross-recognition is due to the fact that the cowpox and smallpox viruses are close evolutionary relatives. Subsequently, this vaccine was so successfully administered in a worldwide campaign that it eradicated smallpox in the late 1970s.

Most infectious diseases arise when a pathogen enters a suitable host (a person, animal, or plant) and begins replicating within the host, often causing disease symptoms in the process. Although viruses and bacteria are the most common pathogens, there are others. For example, malaria results from infection by a protozoan, a microscopic muticelled organism that is considerably more complex than bacteria or viruses.

Pathogens typically infect a particular host species or, in some cases, more than one host species. They do so first by entering the host's body in the water or food the host consumes; in the air the host breathes; through a cut, puncture, or wound in the skin; in the fluids injected through the skin from an insect or animal bite or a contaminated hypodermic needle; or through direct contact of bodily fluids during sexual activity. Once inside the host's body, the pathogen may immediately begin replicating, eventually causing disease symptoms, or it may remain latent for a time lasting days or even years before it begins replicating.

Our immune systems do not act quickly enough to immediately defeat all pathogens, so when a pathogen enters our bodies and begins replicating, we typically suffer for a time from disease symptoms until the immune system ultimately overcomes the pathogen. Some deadly diseases succeed in killing, or permanently damaging, their victims before the immune system wins the battle. Tuberculosis and malaria, for example, are often fatal because they may outcompete the immune system. In spite of its shortcomings, our immune system is very active, attacking a multitude of pathogens throughout our lifetimes. People who are born with a genetically defective immune system often die in infancy from infections their immune systems cannot fight.

Millions of microorganisms enter our bodies every day, but only a tiny fraction of them are pathogens capable of causing disease. When a pathogen enters the body, it recognizes the body as an appropriate host by interacting with a specific chemical signal, such as a particular substance that the host naturally produces as part of its normal metabolism. Pathogens that infect other animals and plants frequently enter our bodies, but they fail to infect us because we do not produce the chemical signals they recognize in the hosts they naturally infect.

Natural selection plays a high-stakes back-and-forth game with infectious disease. People who die from an infectious disease before or during their reproductive years are less likely to leave offspring than are those who escape infection. Thus, any variation in the host's DNA that confers resistance to a deadly pathogen is favored through natural selection, eventually increasing the number of genetically resistant individuals in the host species. However, the pathogen is a living organism with DNA (or RNA, in many viruses) that also carries genetic variations. Those pathogens with a variation in DNA or RNA that overcomes the host's resistance are more likely to survive and reproduce, increasing the proportion of pathogens that can infect previously resistant individuals. As natural selection in the host species favors resistant individuals, natural selection in the pathogen species favors individuals that overcome the resistance. The consequence is accelerated back-and-forth evolution—genes in the host rapidly evolving to confer resistance accompanied by genes in the pathogen rapidly evolving to overcome the resistance, both as a consequence of natural selection.

Evidence of this back-and-forth coevolution of genes in hosts and pathogens is abundant. Nearly all our genes are very similar to their corresponding genes in the chimpanzee genome. However, of the genes that differ most between humans and chimpanzees, a large proportion are those that confer resistance to common infectious diseases. Those genes differ so much because they have been rapidly evolving through accelerated natural selection in our ancestors. If we turn our attention to the pathogens, we see the same pattern. Those genes that allow the pathogen to overcome resistance are highly divergent in closely related pathogen species, evidence that these genes, too, are rapidly evolving.

In chapter 6, we discussed the evolution of resistance to malaria in ancient humans starting about two million years ago, followed by the evolution of a new pathogen species, *Plasmodium falciparum*, which overcame the resistance. The human gene that mutated into a pseudogene conferred very effective resistance that lasted for more than a million years. But eventually mutations in the pathogen allowed it to defeat the host resistance and evolve into an even more serious human pathogen than it once was. It is an excellent, yet tragic, example of the back-and-forth game that natural selection plays.

Might resistance to malaria evolve once again in humans? Actually, resistance is evolving in modern humans in several genes, and it has been doing so for quite a long time. Unfortunately, however, one of the major mutations that confers resistance to malaria in modern humans is not a good one. The mutation resides in the beta-globin gene, which encodes a protein in adult hemoglobin (the same adult gene we discussed in chapter 6). People who inherit just one copy of this mutated gene are resistant to malaria. In areas where malaria is prevalent, people who carry the mutation tend to have more children than those who don't have the mutation, and they pass the mutation on to about half of their children, who are likewise resistant to malaria. But when someone inherits *two* copies of this mutation, one from each parent, he or she inevitably suffers from one of the most serious of human genetic disorders—sickle-cell anemia. This disorder causes red blood cells to assume a sickle shape, instead of the normal rounded shape, and the sickled cells clog the tiny blood vessels called capillaries.

The sickled cells quickly die, causing *anemia*, a severe shortage of red blood cells. The spleen, which removes the dead cells, becomes overloaded and cannot fight infection as it should. The results are devastating. The symptoms—fatigue, widespread pain, swelling, jaundice, eye damage, infections, and fever—begin in infancy and are lifelong. Ultimately, organs throughout the body become so damaged that a slow and painful death often results. Before medical treatments were available, most people with sickle-cell anemia died during childhood. Treatments now allow many who suffer from this disorder to live into late adulthood.

Because natural selection favors people who carry one copy of this mutated gene in places where malaria is prevalent, sickle-cell anemia is most common in those parts of the world and in people whose ancestry traces to those areas. Nowhere is malaria more prevalent than in Africa, and for this reason, sickle-cell anemia is a dreaded genetic disorder there and among people whose ancestors are from Africa. For instance, in Europe and the United States, sickle-cell anemia is the most prevalent genetic disorder among African Europeans and African Americans because natural selection favored resistance to malaria in their ancestors who lived in Africa.

The evidence is clear and unambiguous from a story that spans millions of years of human evolution and is still under way. Human malaria is a consequence of natural selection in humans and natural selection in the parasites, in both modern and ancient times. Moreover, natural selection favoring resistance to malaria explains why sickle-cell anemia affects a significant proportion of modern humans. And this story is just one of many similar stories that tell us how our evolution is inextricably tied to the diseases that afflict us.

THE HUMAN IMMUNODEFICIENCY VIRUS (HIV), WHICH CAUSES AIDS, IS RAPIDLY EVOLVING.

Let's now turn our attention to one of the most feared modern-day viruses—the human immunodeficiency virus (HIV), which causes AIDS. This virus evolved very quickly to become one of the greatest health threats in modern history. It is so dangerous because it attacks the very cells that normally fight viral infections—cells of the human immune system. People who suffer from AIDS typically do not die from AIDS itself but from other infectious diseases that exploit their weakened immune systems.

The first death related to AIDS in the United States happened in 1959, although it was not recognized as AIDS because the disease had yet to be described. Later analysis of the man's tissue confirmed that he was infected with HIV. By compiling molecular information, the infection history, and the geographic distributions of the virus, evolutionary biologists have reconstructed its evolutionary history and determined how it overcame natural immunity in humans and evolved into the distinctly human form known as HIV.[2]

HIV is a retrovirus, which means it uses RNA as its genetic material. Upon infection, it replicates very quickly and easily acquires mutations, resulting in exceptionally high genetic diversity. These two features allow it to rapidly evolve. The virus traces its evolutionary history to related viruses called SIVs (simian immunodeficiency viruses), which infect apes and monkeys. In the heart of Africa, an unknown wild chimpanzee became infected with two different SIVs, one called SIVrcm, which infects red-capped mangabeys (rcm = red-capped mangabey), and another called SIVgsn, which infects greater spot-nosed monkeys (gsn = greater spot-nosed). In the infected chimpanzee—who probably killed and ate these monkeys as chimpanzees sometimes do—the RNA of these two viruses fused to form a new hybrid virus that contained a portion of SIVrcm combined with a portion of SIVgsn. This new virus evolved to become an infective virus in chimpanzees called SIVcpz (cpz = chimpanzee).

The most common form of HIV (called HIV-1 group M) arose when SIVcpz jumped from chimpanzees to humans, probably in the 1920s in the southeast corner of Cameroon. The infection most likely resulted from blood-to-blood transmission when someone who butchered chimpanzees to consume the meat was accidentally cut during the process. From there the virus spread through person-to-person transmission. Infected people carried the virus to the city of Kinshasa in the Democratic Republic of Congo, where in an urban area multiple sexual transmissions allowed the virus to infect a large number of people. There the virus diversified by mutating and then spread to people in the rest of Africa and ultimately the entire world. The highest genetic diversity of HIV-1 group M remains in this part of Africa, evidence that it first evolved there.[3]

To successfully make this jump, the virus had to mutate into a form that could overcome natural immunity to SIV in humans. Mammals have a gene that encodes a protein called tetherin. This protein has evolved to confer resistance to retroviruses by tethering them to the inside of the cell they infect (hence its name), and preventing the viruses from replicating. For SIVcpz to successfully infect a human, it had to overcome the suppressive effect of human tetherin.

Recall that genes conferring resistance to pathogens are among those that differ most in humans and chimpanzees because of the back-and-forth game natural selection plays with genes in hosts and pathogens. The tetherin gene is one of these. The chimpanzee virus, SIVcpz, evolved by acquiring two anti-tetherin genes called *nef* and *vpu*, one from each of the original monkey viruses that fused to form the chimpanzee virus. The *nef* gene mutated to overcome chimpanzee tetherin, but the *vpu* gene remained essentially inert. When the virus jumped to humans, however, human tetherin was so different from chimpanzee tetherin that the *nef* gene could not overcome human tetherin. Instead, the previously latent *vpu* gene mutated to overcome human tetherin, allowing HIV-1 group M to successfully infect humans. This virus has now spread throughout the world and causes 98 percent of all AIDS cases.

A mutation in a second gene, called *gag*, was also required for the chimpanzee virus to jump to humans. Interestingly, a case in which HIV (the human virus) infected a chimpanzee is known, and the *gag* gene of this virus *mutated back* to its original form in the chimpanzee to successfully reinfect its ancestral host, strong evidence that the *gag* gene mutation is essential for human infection.[4]

Group M is by far the most common form of HIV and was probably the first to evolve from chimpanzee-to-human transmission. There are other groups, however, that arose from different mutations and jumped from hosts to humans. They include groups N, O, and P,[5] and a second form of HIV, called HIV-2, which originated when a virus called SIVsmm, found in a monkey called the sooty mangabey, jumped directly from the sooty mangabey to humans.[6] So far, these groups account for just a small fraction of HIV infections. Nonetheless, the available evidence clearly tells us that HIV is still evolving, and new types are appearing at an alarming rate.

HERVs MAY OFFER AN OPPORTUNITY FOR AN AIDS VACCINE.

The human immunodeficiency virus is an example of how a modern virus can evolve to jump hosts and infect people. As we saw in the previous chapter, certain ancient viruses evolved to infect the germline of our distant ancestors. They ultimately became extinct as infectious viruses but evolved within our genome into human endogenous retroviruses (HERVs). As we saw in the previous chapter, over many generations, they replicated and retrotransposed multiple times to the point that there are now more than 443,000 copies of HERVs in our genome. Most HERVs are highly mutated and are often truncated (cut off so that only a piece of the original virus remains), but some of them still carry the ancient viral genes required for infection. Because HERVs no longer propagate themselves through infection, most of the genes they carry are silent and useless, shut off long ago in our evolutionary history.

Because HERVs once were infectious retroviruses, like HIV is now, some of the genes they carry are similar to those in HIV. Using an understanding of how HERVs evolved in our genome, a group of scientists led by Keith Garrison and Douglas Nixon at the University of California–San Francisco, and Brad Jones at the University of Toronto, hypothesized that when HIV infects a cell it might reawaken the previously silent HERV genes in that cell. Their suspicion turned out to be correct—HIV-infected cells produce proteins encoded by HERVs that cells uninfected with HIV do not produce. This discovery offers an innovative possibility for fighting AIDS or even for producing a vaccine against it. If a vaccine can coax the human immune system into attacking the HERV proteins produced by HIV-infected cells, AIDS may be prevented or at least slowed in HIV-positive people. In this case, the ancient evolution of HERVs in our genome may end up helping us combat the modern evolution of HIV.[7]

EVOLUTION EXPLAINS WHY PATHOGENS CAUSE DISEASE SYMPTOMS.

Any variation in a pathogen's genetic material (DNA or RNA) that improves its transmission from one person to another is evolutionarily advantageous for it. This fact helps explain why certain disease symptoms appear with illness. For example, some of the most common infectious diseases cause severe coughing and sneezing—such as the common cold, respiratory infections, influenza, whooping cough, and tuberculosis. Coughing and sneezing are effective means for transmitting pathogens from one person to another, and natural selection should favor pathogens that cause such symptoms.

Other pathogens, such as harmful strains of *E. coli*, *Salmonella*, and *Vibrio cholerae* (the bacterium that causes cholera), produce severe diarrhea in their hosts, which is an effective means for transmitting the pathogens through food and water supplies if proper sanitation is unavailable or neglected. Outbreaks of harmful *E. coli* in the United States and elsewhere are all too frequent when infected food-service workers fail to adequately wash their hands after using the toilet and then handle food products, or when human waste makes its way into water used to irrigate fresh vegetables or fruits. The microbes responsible for these diseases benefit from causing diarrhea because it facilitates their transmission. Thus, natural selection should favor pathogens that produce such symptoms.

Pathogens responsible for sexually transmitted diseases, such as syphilis, gonorrhea, and AIDS, rely on direct sexual contact between people for transmission and have evolved to exploit the tendency of people to seek multiple sexual partners. HIV, for example, thrives in bodily fluids and cannot persist long outside of a host, which is why people do not become infected by HIV from toilet seats, doorknobs, touching, shaking hands, or from coughing and sneezing. However, direct contact of fluids and rupture of small blood vessels during sexual activity allows the virus to readily move from person to person without being exposed to the external environment. Likewise, shared needles for drug injections are a common source of HIV infection because small amounts of contaminated blood may remain in the needle.

EVOLUTION EXPLAINS WHY PATHOGENS COUNTER OUR EFFORTS TO CONTROL THEM.

Fungi (plural of *fungus*) constitute a broad group of organisms that includes molds and mushrooms. They reproduce through microscopic spores that grow and divide when they land on a food source, such as mold growing on stale bread or mushrooms growing on rotting wood. Fungi often grow in the same places as bacteria and must often compete with them for resources. Evolution has given us (along with other vertebrate animals) elaborate and complex immune systems, which attack invading bacteria. However, evolution took a very different course in fungi, giving them the ability to produce chemical substances that are toxic to bacteria. Thus, fungi naturally exclude bacteria where they grow. We call these antibacterial chemical substances *antibiotics*.

Most antibiotics successfully kill certain types of bacteria but are mostly (albeit not entirely) harmless to humans, making them an excellent natural source of medications for treating bacterial infections in people. The first mass-produced antibiotics appeared in the 1940s, and since that time they've saved millions of lives. Diseases that once caused widespread sickness and often prolonged and painful death have been substantially reduced through antibiotic use coupled with the implementation of sanitary conditions. The bacterium that causes plague, for example, is still with us. However, if someone contracts plague, physicians can readily treat it with antibiotics, if caught early enough. The same is true for cholera, typhoid fever, syphilis, gonorrhea, and many other bacterial diseases.

Antibiotics were, and still are, miracle drugs. But because we have been using them on a large scale, we have set bacteria on a predictable course in which natural selection favors the evolution of antibiotic-resistant bacteria. These resistant bacteria are now appearing across the world, rendering previously successful antibiotics ineffective, even useless in some cases.

Some strains of bacteria are now classified as multidrug resistant because over time they have accumulated resistance to several different classes of antibiotics. Multidrug-resistant tuberculosis is an example. The antibiotic streptomycin was introduced in 1944 for treatment of tuberculosis and was soon followed by a series of additional antibiotics for treating tuberculosis.

Researchers quickly recognized that drug-resistant bacteria could arise in people who were treated with a single antibiotic, so multiple-drug treatment became the norm. The widespread use of multidrug treatment throughout the world for tuberculosis was phenomenally successful and resulted in the near eradication of the disease.

However, multidrug-resistant tuberculosis emerged as a serious problem in the 1980s, and it has dramatically increased since that time. For instance, about 10 percent of tuberculosis cases in New York City in 1982 were multidrug resistant, increasing to 23 percent in 1991, and 33 percent in 1993.[8] At the same time, tuberculosis infections began to skyrocket throughout the world.

There are several reasons for the resurgence of tuberculosis; one of the most crucial is the AIDS pandemic. Because AIDS results in a loss of immune function, AIDS patients become more susceptible to a multitude of infectious diseases, which can exploit the weakened immune system. Tuberculosis has become one of the most common diseases to afflict AIDS patients, often with death as the result. By 2007, the number of new tuberculosis cases worldwide had escalated to more than nine million, with more than one million deaths, making tuberculosis the world's most deadly bacterial disease.[9] As millions of people are treated for tuberculosis with multidrug therapy, multidrug resistance is repeatedly appearing as a consequence of natural selection. Scientists are now concerned about a possible worldwide epidemic of multidrug-resistant tuberculosis.[10]

Unfortunately, hospitals—where people with serious infectious diseases are often treated—have become breeding grounds for multidrug-resistant bacteria. Most hospitals take precautions to control bacteria through careful and thorough disinfection. But the presence of resistant strains of disease-causing bacteria has prompted physicians to turn to newer, more powerful antibiotics to combat antibiotic-resistant bacteria. As is happening with tuberculosis, the effectiveness of even these powerful antibiotics is waning, as bacteria acquire multidrug resistance. Decades ago, evolutionary biologists warned that multidrug-resistant bacteria would appear. The warning has proven true.

Antibiotics are also routinely used in modern animal production. Feed lots, large dairy operations, and crowded poultry cages are prime sites for bacterial infection because animal-to-animal transmission of pathogens can be rampant where so many animals are in close contact. To prevent outbreaks, animal producers routinely supplement animal feed with antibiotics. Such a situation creates the very conditions needed for natural selection to favor antibiotic resistance, which is now a reality in animal production.

Moreover, certain bacteria that cause human disease, such as harmful *E. coli*, may thrive in cattle, pigs, sheep, and poultry without causing disease in these animals. The outcome of widespread antibiotic use in animals is natural selection favoring antibiotic resistance in these bacteria that cause human disease, threatening our ability to treat infected people. Although biologists have warned that the outcome of excessive antibiotic use in animal production is a large-scale threat to human health, it remains a widespread practice.

EVOLUTION EXPLAINS WHY CERTAIN SPECIES NATURALLY PRODUCE MEDICINES AND DRUGS.

As we saw earlier, fungi naturally produce antibiotics to protect themselves from bacterial competition. Plants also naturally produce many of the drugs, medications, and nutritional supplements people use for medicine and improved health, as well as some of the most potent and dangerous substances people abuse when addicted to drugs. They include nervous system stimulants (such as cocaine, nicotine, and caffeine), nervous system inhibitors (atropine and scopolamine are examples), pain relievers (including morphine, codeine, and aspirin), antioxidants (such as vitamins A, C, and E), and cell-division inhibitors (for instance, paclitaxel, vincristine, and vinblastine).

People often wonder why plants naturally produce substances that are medically beneficial to humans, or are harmful addictive drugs. As it turns out, these substances provide benefits to plants to such an extent that natural selection should favor the evolution of their production. Antioxidants protect plants from the damaging effects of chemicals produced during photosynthesis as the plants are exposed to the sun, and they provide similar protections for us. Our bodies also produce natural antioxidants, but we are dependent on plants for certain types, such as vitamins A and C. There is strong evidence that our distant ancestors' bodies naturally produced vitamin C, as most animals do.

However, during primate evolution, an essential gene for vitamin C synthesis mutated into a pseudogene (called *GULOP*), which we now share with our primate relatives. Natural selection did not disfavor this pseudogene because it had become redundant—vitamin C was so prevalent in the plants consumed by early primates that the gene had become superfluous. Now we, and our primate relatives, must consume vitamin C in our diets, usually from plant sources, to compensate for our evolved inability to naturally produce it.[11]

Many of the chemicals produced in plants have powerful effects on nervous system activity, and plants have no nervous systems. For example, what evolutionary advantage do cacao (chocolate), coffee, and tea plants naturally derive when they produce caffeine, a natural neural stimulant? The answer is clear when viewed in terms of natural selection. Like fungi, plants do not have elaborate immune systems resembling ours but instead produce a diverse array of chemicals that inhibit or kill pathogens and pests. These chemical compounds are called secondary metabolites, *secondary* because they are not directly essential for development or reproduction. Instead, they confer a survival advantage through a secondary role, such as protecting plants from pathogens and pests. Many of these secondary metabolites in plants attack insect nervous systems. Our nervous system functions in a manner similar to insect nervous systems (because of an ancient common evolutionary origin of nervous systems), and these substances affect our nervous system in ways that are similar to their effect on an insect's nervous system. Caffeine, for example, protects coffee plants from insect predation by overstimulating the nervous systems of insects who feed on coffee plants. Obviously, our bodies are considerably larger than insect bodies. A dose of a secondary metabolite in a plant that kills an insect eating the plant may be quite safe for us, thanks to our body size. These compounds, while not killing us, can still affect our nervous systems, stimulating or suppressing it in various ways.

For much of human history, people have sought medicines in plants, and much of what we have learned about how drugs work comes from studies on the evolution of plant-secondary metabolites and their effect on humans. This understanding has allowed chemists to take things a few steps further by starting with a plant-secondary metabolite and chemically modifying it to make it more effective or to remove harmful side effects. In these cases, a product of evolution is the starting point for improvement through human ingenuity.

While chemists modify secondary metabolites to improve them, a dedicated group of botanists are searching the world for plants that contain previously unknown secondary metabolites that may prove useful in medicine. Among these researchers are ethnobotanists who study the medicinal practices of native cultures throughout the world to determine what plants they use and how they use them. Back in the laboratory, they extract secondary metabolites from these plants to determine which ones have medicinal properties.

VACCINES EFFECTIVELY PREVENT VIRAL INFECTIONS, BUT VIRUSES MAY RAPIDLY EVOLVE.

Antibiotics save thousands of lives through treatment and prevention of bacterial infection. Some of the most prevalent infectious diseases, however, are caused by viruses, and viruses are mostly immune to antibiotics. A few antiviral drugs, such as those that treat AIDS or influenza, are now available, and more are likely to appear in the future. However, for most viral infections, we must rely on our immune systems to combat the infection.

In the late 1800s, a group of visionary scientists surmised that they could possibly trick the immune system into fighting a particular disease when the disease was not really there. People whose immune system fights a false infection produce antibodies that attack both the false and the real infection, thus preventing the disease. The false infection is usually a killed or weakened virus administered as a vaccine. When people are vaccinated, the vaccine is unable to cause the disease, but the immune system still attacks the vaccine, rendering the person immune to the real disease. When viruses of the real disease enter immunized people, antibodies immediately attack the invading viruses, marking them for destruction by the immune system before the viruses have a chance to multiply.

The ability of viruses to overcome immunity in vaccinated people varies widely. Some viruses, such as those that cause measles, mumps, chicken pox, whooping cough, and smallpox, do not evolve rapidly, so immunity acquired through vaccination lasts for years, in some cases even a lifetime. Physicians often recommend booster vaccines for some of these diseases, not because the virus has evolved but because the immunity may wane over time.

Other viruses, however, evolve very quickly, and people must be revaccinated against new strains as they appear. The best known are the influenza viruses. Each year, scientists must develop new vaccines against rapidly evolving strains of seasonal influenza—we usually call those vaccines "flu shots." On occasion, especially dangerous strains of the flu appear, such as the H1N1 strain, popularly known as swine flu. The H1N1 strain threatened to cause an outbreak in 2009 and 2010 and, to a lesser degree, in 2011. Massive production and widespread administration of H1N1 vaccine contributed to a much less serious outbreak of the disease than might have occurred without vaccination.[12]

NONINFECTIOUS DISEASE IS ALSO A CONSEQUENCE OF EVOLUTION.

A large number of diseases are noninfectious—they result from something other than an infectious microorganism. Cancer is perhaps the most common and most feared noninfectious disease. It is true that a few cancers result from viral infections, such as Rous sarcoma and certain types of cervical cancer, but the vast majority of cancers are noninfectious.

The initial stages of cancer begin when a mutation happens in one of a large number of so-called cancer genes in our cells. A mutation in one of these genes is not enough to cause cancer. However, a second gene may mutate, then eventually a third, and perhaps a fourth. After several of these genes have mutated in a cell, the cell may begin dividing when it should not. A mass of mutated cells, resulting from the original mutated cell, eventually develops into a tumor. In later stages of cancer, cells break away from the original tumor and implant themselves elsewhere in the body, where they, too, grow into tumors, causing the cancer to spread.

Some people wonder why we carry genes that cause cancer. Shouldn't natural selection have removed them long ago? The answer, it turns out, is quite simple. Many of the genes responsible for cancer are genes that are absolutely essential for us because they regulate cell division. When an egg cell in your mother united with a sperm cell from your father, you were briefly a single-celled organism. At that point, genes that regulate cell division stimulated that one cell to divide into two, and those two into four, and those four into eight, and so on, until eventually you became an embryo, fetus, infant, child, and adult. At various points during your biological development, the genes that stimulate cell division in various organs were turned off, and your cells there ceased dividing. Although some of your cells continue to divide, such as the cells that line your intestinal tract, the vast majority of your cells will never divide again. They are destined to carry out their own particular roles until they (or you) die. However, when mutations disrupt the normal controls that suppress cell division, cells start dividing again when they should not, and cancer can be the result.

The chance of any one cell accumulating enough mutations to become cancerous is very small. However, given enough time and opportunities, even very unlikely events can happen. The human body contains as many as thirty trillion cells, most of which are capable of accumulating mutations. So if you live long enough, the chance that at least one of your cells will accumulate enough mutations to become cancerous is quite high. For this reason, the probability of cancer increases with age.

The fact that modern interventions in sanitation and healthcare have reduced infectious disease also explains why cancer is more prevalent now than in the past. As infectious disease is reduced, life expectancy rises, and cancer increasingly becomes a cause of death as people who might have previously died from infectious disease grow old and eventually succumb to cancer.

Moreover, radiation and certain chemical substances called carcinogens can increase the incidence of cancer. They do so by increasing the rate of mutations. Ultraviolet light, present in sunlight, for example, damages the DNA in our skin cells and can cause mutations responsible for skin cancers. Radioactivity from nuclear waste, detonated nuclear weapons, even X-rays, natural radon gas, and radioactivity from space (to which we all are constantly exposed) can cause cancer. The doses of natural radiation from space and diagnostic X-rays used in medicine and dentistry are typically low enough that they do not substantially increase cancer risks, however.

All types of smoke contain polycyclic aromatic hydrocarbons (PAHs), a potent class of carcinogens. Tobacco smoke is concentrated when inhaled during cigarette smoking and exposes lung tissues to high levels of PAHs, which cause mutations in lung cells that can eventually lead to lung cancer. Polycyclic aromatic hydrocarbons are also present in automobile exhaust, smokestack emissions, secondhand tobacco smoke, even in the smoke from a backyard barbeque or campfire. Although not as concentrated as in tobacco smoke, long-term repeated exposure to the carcinogens in these sources can also cause mutations that result in cancer.

There is abundant evidence that evolution through mutation and natural selection has shaped a large suite of genes whose principal role is to stop cancer before it starts. Hundreds of studies have examined the evolution of these genes. We'll look here at just one of them—the *P53* gene—as an example, because it is one of the most important genes we have for cancer prevention.

Most of your cells have the surprising ability to destroy themselves if they begin to threaten your health through a process of programmed cell suicide called *apoptosis*. The genes that trigger apoptosis remain silent in most cells, activating themselves only when a cell is so damaged that it threatens the body. As long as the genes governing apoptosis are intact and free of debilitating mutations, a cell that has accumulated enough mutations to start down the road toward cancer will kill itself before cancer can develop. Many of our cells do this without us ever realizing it, and we escape what could potentially be early death from cancer.

However, the genes that trigger apoptosis are also subject to mutation, and when they mutate, a cell's ability to kill itself may be lost. One of the most important of these cell-suicide genes is called *P53*, a master gene whose role is to regulate other genes that trigger apoptosis. In essentially all cancerous cells, either the *P53* gene or one of its related genes must have acquired a mutation, rendering it nonfunctional. Without these mutations, cells on the road toward cancer enter apoptosis and self-destruct, protecting the organism from cancer.

There is powerful evidence, derived very recently from comparison of genomes and studies of gene function, of how *P53* and its related genes evolved.[13] Its ancestral gene is not present in bacteria, yeast, or plants but is present in the genomes of single-celled choanoflagellates, which evolutionarily are the closest single-celled relatives of animals. The ancestral gene of *P53* must have arisen at least one billion years ago in early single-celled ancestors of animals.

The ancestral gene has remained in most invertebrate animals, including worms, insects, sea anemones, sea urchins, and clams. In these organisms, its role is to promote apoptosis, not in cancer cells but in damaged egg and sperm cells, thus preserving DNA integrity in offspring. Even in its earliest evolutionary stages, this gene promoted apoptosis.[14]

In early ancestors of vertebrates, the ancestral gene duplicated once to form two copies, one of which diverged into *P53*. In early vertebrates, about the time bony fish emerged, the other copy duplicated again, diverging into two genes called *P63* and *P73*. Most vertebrates now have copies of all three genes—*P53*, *P63*, and *P73*.

During the earliest stages of vertebrate evolution, when all vertebrates were fish, *P53* first began to regulate apoptosis in body cells, as opposed to egg and sperm cells, thus preventing cancers in early vertebrates. The reasons for this are clear. Adult invertebrates, such as insects and worms, have very few cells that are capable of dividing—by the time the adult has emerged, nearly all the cells are present, and the adults have short lifetimes, not enough time for cancer to prohibit reproduction. Vertebrates, by contrast, have longer life cycles and have cells that are capable of dividing and repairing damaged body parts. They are more prone to cancer, and natural selection should favor any process that prevents or delays cancer. In this case, evolution co-opted a gene that originally protected egg and sperm cells and diverted its function to the rest of the body.[15]

THERE ARE MANY OTHER TYPES OF NONINFECTIOUS DISEASES, AND THEY, TOO, ARE SUBJECT TO EVOLUTION.

Genetic disorders, such as sickle-cell anemia and cystic fibrosis, are the result of mutations in single genes that disable those genes and eliminate their function. People who inherit mutated copies of these genes from both parents suffer from these genetic disorders because both copies of the gene are disabled. Other genetic disorders are complex, the result of variations in numerous genes, each of which has a minor effect. But when particular genetic variants in these several genes are combined through inheritance, a genetic disorder appears.

Genetic disorders are typically rare because the mutations that cause them are gradually eliminated through natural selection. They persist, however, because natural selection works slowly at removing them, often requiring many generations. By then, new mutations in the gene have appeared in other individuals, and these new mutations are likewise subject to slow and gradual removal by negative purifying selection.

In some cases, devastating genetic disorders may persist because their effects appear after people who carry the mutation have reproduced. Huntington disease, for example, is a tragic genetic disorder that causes gradual brain and nerve deterioration and ultimately death. Many people who carry the mutated gene lead normal lives with no symptoms until their late thirties to early forties. The symptoms begin with gradually increasing nerve and brain deterioration until people afflicted with the disease lose all motor control and mental function. Most die from the disorder before age fifty. Natural selection has little effect because most people who carry the mutations have reproduced before symptoms appear. On average, half of their children carry the mutation, and they pass it on to half of their children, who pass it on to half of their children, and so on, generation after generation.

Autoimmunity is another cause of noninfectious disease. In this type of disease, the immune system mistakenly recognizes substances that are naturally part of a person's body and attacks them as if they were foreign substances. For instance, in rheumatoid arthritis, the body's immune system mistakenly attacks tissues in the joints as if these tissues were infectious viruses or bacteria. The membranes, cartilage, and bone in the joints are incessantly inflamed, and they gradually deteriorate because of the immune-system onslaught. Severe cases result in bone disfiguration. Susceptibility to the disease can be inherited (it has appeared in several generations in my own family), but because the first symptoms typically appear after age forty, natural selection has little effect on removing the mutated genes that confer susceptibility to it.

Hyperthyroidism is another autoimmune disease that appears when a person's immune system produces antibodies against receptors in the thyroid gland that typically respond to a hormone called thyroid stimulating hormone. The mistakenly produced antibodies attach themselves to the receptors and overstimulate the thyroid gland, causing it to produce too much thyroid hormone.

Autoimmunity illustrates the imperfect nature of evolution. Research conducted over the past fifty years or so has unraveled much of the mystery of how our immune system evolved. The system is effective and keeps every person on the planet (as well as every other mammal on the planet) alive—we would all die from infections during infancy were it not for our immune systems. But our immune system has a complex and haphazard organization that can be readily explained if it arose through a complex evolutionary process, by preservation of fortuitous mutations and DNA recombination over tens of millions of years. Its haphazard organization, unfortunately, leaves it prone to mistakes, and autoimmunity is one of the major types of mistakes it is capable of making.

EVOLUTIONARY GENOMICS IS RAPIDLY ADVANCING MEDICINE.

The birth of the twenty-first century welcomed advances that are beginning to revolutionize modern medicine. As we've seen, both simple and complex genetic disorders are a consequence of mutations in DNA. However, it is becoming increasingly clear that most human diseases and ailments have a genetic component, such as inherited susceptibility to infectious disease, autoimmune disease, or cancer.

Humans have a considerably higher risk for cancer when compared with our great ape cousins. Some of this higher risk can be attributed to our longer life spans, but even when the effect of life span is accounted for, humans appear to have an inherently higher risk conferred by our genetic constitution and recent evolutionary background. In particular, some of the most common cancers in humans, such as breast, lung, and prostate cancer, which constitute about 20 percent of all human cancers, are extremely rare in nonhuman primates.[16] Certain genetic situations unique to humans may predispose us to cancer. For example, in the previous chapter, we discussed the *NANOGP8* retropseudogene, which arose exclusively in the human ancestral lineage since its ancestral divergence from the chimpanzee ancestral lineage. Recent research has shown that this retropseudogene is not entirely inactive but is turned on as a gene (called a retrogene) in some cancerous cells, and that its protein product promotes cancer.[17]

The prevalence of cancer in humans has led scientists to speculate that natural selection may be favoring recent mutations that confer resistance to cancer in humans, and genome-wide surveys offer the possibility of identifying variants that protect against cancer. Because genome-wide comparisons have been possible for only a few years, and because the genetic basis of cancer is complex, research in this area is still in its infancy. Nonetheless, evidence of an evolutionary role for cancer in some important genes is emerging.

Scientists at the Universidad de Oviedo and the Universitat Pompeu Fabra in Spain and the University of Washington in the United States compared 333 genes known to be related to cancer in the human and chimpanzee genomes.[18] Evidence of recent evolution in humans related to cancer points to two heavily studied genes: P53 and BRCA1.

As we saw earlier, P53 is a master gene that governs initiation of apoptosis—it protects the body against cancer by causing cancerous cells to self-destruct and is mutated in the majority of cancer cases in humans. This gene is highly similar in all primates, and many people carry essentially the same version as in other primates. However, a large proportion of humans throughout the world carry a mutated variant version of this gene, which has been extensively studied by scientists throughout the world. This variant appears to enhance P53's ability to induce apoptosis and protect against certain types of cancer early in life for people who carry it, evidence that natural selection may be favoring this variant in modern human populations.[19]

The second gene, BRCA1, is one of several genes known to be associated with susceptibility to breast cancer in humans. The gene itself has a protective effect against breast cancer, and people who inherit mutations in it have a higher risk of breast cancer. The DNA surrounding this gene bears the signature of a selective sweep in humans, indicating that natural selection has favored this gene in human evolutionary history.[20] A DNA test can identify people who carry mutations in this gene and thus are genetically predisposed to breast cancer. If identified early, many cases of breast cancer can be successfully treated. A DNA test can identify people who are more susceptible to breast cancer so they can be more intensively monitored for early detection, so that if breast cancer appears, treatment can be immediately implemented during the earliest stages. The economic benefit for genetic testing is so great that this gene has made headlines because of a high-profile patent dispute over genetic testing for breast cancer risk.[21]

Recall from previous chapters that a selective sweep is a rapid increase in a particular variant over a short period of time through intensive natural selection. Selective sweeps often leave a telltale signature in the DNA. By looking for such a signature, scientists can identify genes that have been favored by natural selection during recent human evolution. The observation that a gene variant underwent a selective sweep indicates that it is a strong candidate for genetic testing to identify people who are more prone to certain types of cancer, genetic disorders, or diseases. For instance, an international team consisting of scientists at institutions in the United Kingdom, Germany, and the United States conducted a genome-wide survey of 142 genes known to interact with the human *P53* gene, looking for evidence of natural selection. They found a signal of a selective sweep favoring several recent mutations in the *PPP2R5E* gene.[22] Studies such as this are an important first step for identifying genes that are important in cancer research, and for identifying people who are at risk for certain types of cancer so they can be closely monitored for early detection and successful treatment.

And it's not just cancer. Thousands of variants in human DNA, which are a consequence of recent human evolution, are known to be associated with genetic disorders and susceptibility to many types of disease. A single DNA test is now available that can simultaneously detect thousands of these variants, and the cost of such a test is dropping precipitously to the point that large-scale genetic testing of people may become routine. Currently, a physician's choice of which medication is most appropriate for treating or preventing a disease rarely takes genetic background into account. With large-scale genetic testing, it is now possible for medical researchers to identify which treatments are most appropriate for people who have a particular genetic background, thus tailoring treatments to the people who will most benefit from them. For example, recall our discussion in chapter 5 regarding telomeres, which tend to erode in our cells as we age. A genetic test for telomere length in cells that circulate in the blood can identify people who, because of their genetic background, are likely to benefit from early treatment with statin drugs to prevent or delay the onset of heart disease.[23]

People are often concerned that genetic testing may pose a serious risk, not to their health but rather their economic well-being. Given that many people receive health insurance through their employers, might a health insurer or employer discriminate on the basis of a genetic test? Genetic discrimination with regard to employment and health insurance has happened in the past and is the stuff of science fiction novels and movies about the not-so-distant future.[24] On the other hand, genetic testing may spare insurance companies and employers large healthcare expenses by identifying risks and averting serious health issues through early detection, effective treatment, and prevention. To prevent discrimination in the United States, the Genetic Information Nondiscrimination Act, known as GINA, which prohibits discrimination on the basis of genetic constitution, was signed into law in 2008.[25]

IS MODERN MEDICINE DEFEATING HUMAN EVOLUTION?

There is no doubt that modern medicine has allowed people to live and reproduce who, in the past, would have died before reproducing. I am one of them. At age two, I would have died, had modern surgery not been available. Medical intervention has unquestionably reduced the effects of natural selection in humans. People with poor eyesight who would have struggled to survive as hunters on the African savannah now are able to enjoy good vision thanks to corrective lenses and LASIK surgery. The genetic disorder phenylketonuria (PKU) at one time caused such severe mental retardation that most who had it were institutionalized and did not reproduce. Now, a simple and effective treatment entirely prevents mental retardation, so people with PKU live relatively normal lives and on average have as many children as those who do not have it. The list of such examples is very long.

So is modern medicine destroying our species by reducing the effect of natural selection and allowing detrimental mutations to persist? The answer is not a simple one. As a species, we have relatively few progeny per mating pair and relatively long generation times. As a consequence, natural selection in humans nearly always operates very slowly, requiring multiple generations and long periods of time to have any appreciable effect. Modern medicine has been with us for so short a time, and for so few generations, that its effect on the evolution of our species has thus far been negligible. In the long term, over many generations, it certainly will have an effect.

But we cannot reliably predict what that effect will be. We can expect scientists to continue making discoveries in medicine that we can now hardly imagine, just as our great-grandparents could not have predicted many of the astounding medical advances we now have. We can also be certain that new infectious diseases that never before afflicted our species will evolve—AIDS is a modern example. Likewise, old diseases once nearly defeated, such as tuberculosis, may evolve to overcome our defenses against them, and biomedical researchers will seek ways to combat them. One thing is certain. The power of natural selection to affect human evolution is slow enough that we need not be overly concerned about medicine having an appreciable effect on human evolution in our lifetimes, or in the lifetimes of our descendants for several generations to come. The effect, although certain, will be very slow and gradual, and any appreciable impact will come when medicine is very different than it is now.

The same cannot be said for the pathogens that infect us. Most of them are capable of rapid evolution. Some can pass through several generations per day, and millions of cells can be derived from a single cell over a very short period of time. Modern medicine *is* having an immediate effect on their evolution, one that scientists and healthcare professionals must constantly monitor and address.

WHY DOES IT MATTER?

We live in an unprecedented time in human history. Our understanding of human evolution is opening doors to a much more refined form of medicine that will increasingly take into account the genetic background of people. Early detection and treatment of cancer and other diseases owe their increased effectiveness to our dramatically improved understanding of how our genomes are evolving and what it means for our health. We are likely to see increased reliance on genetic testing based on evolutionary theory in the future.

The worldwide human population is larger by orders of magnitude than at any other time in the five-million-year history of hominins and the two-hundred-thousand-year history of our species. Moreover, we are far more mobile than at any other time in our history. People whose ancestors were once geographically and culturally isolated now interact routinely. As I write these words, I am flying on an airplane over the southern tip of Greenland, returning to the United States from a trip to Europe where I traveled through nine countries with eight different official languages.[26] These different languages evolved because of geographic and cultural isolation, which also produced a degree of genetic isolation that ultimately influences the biological evolution of our species. That isolation is now breaking down in just a few generations, a phenomenon welcomed by some and lamented by others. From a scientific viewpoint, decreasing isolation should be beneficial to us as a species because it promotes greater genetic diversity through recombination of genes that would not happen if cultures remained isolated.

From an evolutionary perspective, we have some immediate and serious concerns for our future. Some of these, such as world population growth and feeding our burgeoning population, are topics for upcoming chapters. Regarding our health, three of the most urgent challenges are the rapid spread of pathogens, accelerated evolution of new pathogens, and the alarming emergence of multidrug-resistant pathogens.

Our mobile society allows pathogens that in the past would have remained restricted to a particular area to spread throughout the world. Such spread is not unique to modern times. The spread of plague from the fourteenth through nineteenth centuries was a consequence of rats and fleas inadvertently carried by humans as part of international trade. The current AIDS pandemic is a modern example of how a disease can rapidly spread. The scientific evidence tells us that the virus has independently jumped at least twelve times from other primates to humans, although only one of these viruses (HIV-1 group M) has spread widely. Some have questioned why the virus is jumping now and not in the past. In reality, it probably *did* jump in the past, but infections remained localized, did not spread widely, and went undetected until they eventually became extinct. According to Paul Sharp and Beatrice Hahn, two of the leading researchers on the evolution of HIV:

The opportunities for chimpanzee- or monkey-to-human host jumps [of viruses] have existed for hundreds or thousands of years, and it must be expected that many such transmissions occurred in the past. However, only in the twentieth century did such viruses spread to detectable levels in the human population. In west central Africa during the early part of that century, the destabilization of social structures by invading colonial powers, the origin and rapid growth of major conurbations and the widespread use of injections may all have contributed to provide an unprecedented opportunity for dissemination of blood-borne viruses.[27]

Scientific research that allows us to quickly identify how new pathogens evolve permits us to develop better methods for reducing their spread and for treating the diseases they cause. The H1N1 influenza virus is an example. By quickly identifying it, fully sequencing its RNA, and tracing its evolution, scientists were able to develop a vaccine within a short period of time, and healthcare professionals had the information and tools they needed to target their efforts to effectively contain its spread. What could have been a worldwide pandemic, causing millions of deaths, instead was a relatively minor outbreak.

In spite of multidrug-resistant bacteria, antibiotics remain an important tool for treating and controlling bacterial infections. Physicians and other healthcare professionals have already initiated a worldwide campaign to slow the development of drug-resistant bacteria. Efforts to find a wide range of new antibiotics are proving successful but cannot be the whole answer. Many physicians, because of evolutionary research and education, now judiciously limit how often they prescribe these new antibiotics to restrict the opportunities for resistant bacteria to arise through natural selection.

A large-scale campaign aimed at educating physicians about their role in limiting the emergence and spread of drug-resistant bacteria has also seen considerable success. A major problem, however, is public education. When people contract bacterial infections treatable with antibiotics, they must continue the full course of antibiotic treatment prescribed by the physician even when symptoms disappear. Doing so helps to ensure that all the bacteria are eradicated from their bodies, thus preventing antibiotic-resistant types from evolving. In spite of warnings from physicians and written instructions that come with antibiotics, people all too often cease taking their antibiotics as soon as symptoms disappear, instead of taking the full course as prescribed. By doing so, they unwittingly endanger themselves and the entire public by contributing to the emergence of drug-resistant bacteria.

Campaigns to broadly immunize children have produced astounding results. The spread of certain deadly viruses has been stymied, and in some places nearly eliminated. These campaigns, however, can be a victim of their own success. When people no longer see firsthand the devastating effects of diseases like whooping cough or polio, they often choose to not immunize themselves or their children. Unfounded fears that vaccines may cause autism or other disorders exacerbate the issue. As the parent of a severely autistic child, I understand the fear firsthand. However, the well-founded risk of allowing viral diseases to resurge and evolve far outweighs the unfounded fear that immunizations cause autism, which creates an overall threat to public health. Because so many children have not been immunized in recent years, some of these viral diseases are reemerging as threats. If they gain a foothold, they will evolve more rapidly to overcome our efforts to control them.

Antibiotics and immunizations are among several tools in the arsenal against the spread of pathogens. Unsanitary conditions contribute more to the spread of most pathogens than any other factor. Simple acts, such as washing hands before meals and after using the toilet can have a major impact on preventing the spread of pathogens. Alcohol-based hand sanitizers are effective, as are antibacterial cleansers used in homes, workplaces, and especially healthcare facilities, although there is the possibility that with widespread use they, too, may lose effectiveness because of natural selection. Some of the most important workers in hospitals and clinics are those who keep the facilities clean and sanitary. Their efforts, though largely unnoticed, save the lives of countless people. Those who handle food in grocery stores, food-processing plants, cafeterias, and restaurants must likewise observe sanitary practices to prevent the spread of pathogens through contaminated food.

Much of the world is afflicted with an appalling lack of sanitation. Both urban and rural areas have unsafe drinking water, inadequate or nonexistent sewage disposal, poor garbage management, and contaminated food. Such conditions affect not only the impoverished people who are forced to live under such conditions but also the entire world. These places serve as breeding grounds for pathogens, which can be easily spread to areas where sanitation is more advanced. In many cases, modest investments in sanitation, such as providing free bottled drinking water or water purification tablets, improved public health education, and implementation of basic sewage and garbage disposal infrastructure could do much to prevent the spread of infectious disease. Unfortunately, public funding and foreign aid are often inadequate, or are diverted to less important priorities for political reasons.

Scientific research that unravels the mysteries of how we and the pathogens that infect us have evolved and are evolving is one of the most crucial aspects of improving overall human health. The knowledge we gain through such research has given us the means to detect, prevent, and treat disease. Understanding in detail how pathogens evolve helps scientists develop more effective drugs to combat disease and vaccines to immunize us against them. Public health professionals rely on information derived from evolutionary studies to optimize efforts to control the spread of pathogens. Unraveling the evolutionary basis of noninfectious disease has allowed scientists to make remarkable progress in early detection of cancer, autoimmune diseases, and genetic disorders, and in developing effective treatments for these diseases.

Our health is ultimately a consequence of our evolutionary history and the evolution of the organisms that interact with us. These organisms include the pathogens that infect us, and in this chapter we've barely scratched the surface of our vast and intricate evolutionary interaction with them. Beyond these pathogens, some of the most important organisms that coevolve with us are the plants and animals we use for food. We have radically shaped their evolution during a relatively small tick on the evolutionary clock, and they in return have impacted our own evolution. The next chapter tells their story.

CHAPTER 9
EVOLUTION AND OUR FOOD

My maternal grandparents lived in a rural town where they oper-
ated a small family farm. My great-grandparents were among
the town's early settlers, having moved there in the late nineteenth century.
But they were not the first to settle there. From about 800 to 1150 CE, the
Ancient Pueblo People (also known as the Anasazi)[1] occupied the site, leaving
the remains of mound homes and countless artifacts, including pottery,
arrowheads, and seeds. Much of the ancient settlement is now covered by the
modern town.

Relics of this ancient occupation were easy to find. Not far from the edge
of town is a flat volcanic boulder, about the size of a small table. Carved into its
surface are a series of circular depressions. We called it the game rock, which
is possibly what it was (figure 9.1). In the nearby mountains are cliff murals
of magnificent rock art, which tourists often view. Just outside of town is a

Figure 9.1. The "game rock" carved by the Ancient Pueblo People.

large burial site where, for generations, the Ancient Pueblo People interred their dead with their treasures. Professional anthropologists excavated the site during the 1950s, removing important artifacts and cataloging them for museum collections.

The Ancient Pueblo People were farmers who cultivated a type of corn that was probably well adapted to the arid environment where they lived. Remnants of their corn were scattered among the ruins they left behind. The corn they grew, however, was not native to the region. For millennia, the range of cultivated corn expanded northward from its evolutionary origin in Mexico, as people carried it with them during their migrations to new areas, until it reached the Ancient Pueblo People.

I found myself as a small boy learning from my grandparents to grow corn on the same land where these ancient people had planted it centuries earlier. Among the many lessons I learned, one in particular has returned to my memory time and again. It was early evening after a hot summer day, just before supper, and my grandfather and I were in the garden picking sweet corn. I noticed a tall corn plant with an especially large ear. Eagerly, I went to pick it, only to have my grandfather stay my hand before I could touch it. "That's the best ear of corn in the garden," he said. "We save the best ears for next year's seed."

My grandfather was practicing a tradition that the Ancient Pueblo People before him also probably practiced, one dating back to the earliest roots of human civilization. Plant breeders now call this age-old tradition *mass selection*—allowing plants to naturally breed, then selecting seed from the best plants for the next generation. The concept is simple, but the long-term outcome is profound. For thousands of years, people have known that plants and animals pass on their traits to offspring through inheritance. It made sense, therefore, for them to save the best plants and animals as breeding stock for the next generation and, by so doing, continually improving the genetic quality of their food.

OUR EVOLUTION IS INEXTRICABLY TIED TO THE EVOLUTION OF DOMESTICATED PLANTS AND ANIMALS.

By about ten thousand years ago, modern humans had migrated to all major regions of the world they now occupy. As human populations grew in numbers, the demand on local resources intensified and people were gradually less able to support themselves exclusively through hunting and gathering their food in nature. Modern archaeology and genetics have given us a fairly detailed picture about the next steps. Instead of relying entirely on nature to provide their food, people began intentionally cultivating wild plants and caring for wild animals, practicing mass selection among them to gradually obtain those best suited to their particular conditions. Over a few thousand years, the power of their intentional selection shaped the evolution of new domesticated species, which provide most of the food we now consume.

Since Darwin's day, this form of human-directed mass selection has been called *artificial selection* to distinguish it from *natural selection*. Just as natural selection ultimately results in the evolution of new species, artificial selection has produced new species, even more rapidly in most cases than natural selection. These new species that arose through artificial selection are the domesticated plant and animal species we now use for food. Wheat, corn, rice, potatoes, barley, oats, beans, soybeans, vegetables, fruits, cotton, flax, cattle, chickens, sheep, goats, pigs, turkeys, and many others are newly evolved species of plants and animals domesticated by humans, mostly during the past ten thousand years.

This process of domestication has continued to the present through mass selection and has accelerated since the development of modern plant and animal breeding. Beyond the plants and animals our ancestors domesticated for food and fiber, they also domesticated ornamental plants we use to beautify our surroundings, and animals we treasure as pets.

In most cases, we can trace the origin of each domesticated species to a particular place in the world, and often we find its wild relatives there. Corn, for example, was domesticated in central Mexico from a grass known as teosinte, which still grows wild in the region. Corn and teosinte have many

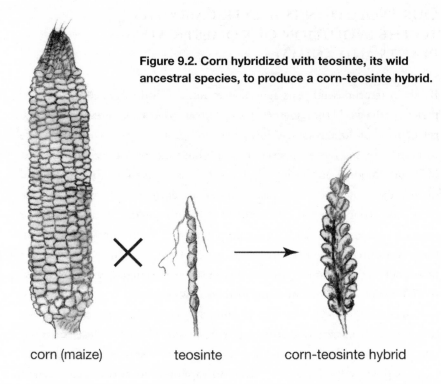

Figure 9.2. Corn hybridized with teosinte, its wild ancestral species, to produce a corn-teosinte hybrid.

corn (maize)   teosinte   corn-teosinte hybrid

similarities and are almost undistinguishable as seedlings. However, as shown in figure 9.2, teosinte has a relatively small seed head, with a few triangular-shaped seeds lined up in a single row. When the plant matures in the fall, it releases the seeds onto the ground where they sprout the following year to regenerate the species.

Corn, by contrast, has an ear with multiple rows of seeds that adhere tightly to it. This adherence of the seeds is beneficial to humans but disastrous for the plant if it must survive without human intervention. People can readily harvest corn seeds, and keep the seeds for themselves, because the seeds remain on the plant. But corn, on its own, has no means to naturally disperse its seeds. Instead, people must intentionally remove the seeds from the ear and plant them in the soil. If we were to suddenly go extinct, corn would go extinct along with us.

Corn is still so closely related to teosinte that the two can hybridize and produce fertile offspring that are intermediates between their two parental species (figure 9.2, page 226). If these hybrids are allowed to self-fertilize or mate with each other, they do not produce offspring like themselves but instead yield a wide array of types ranging from those resembling teosinte to those resembling corn, and many types in between.

The domestication of corn from teosinte was a long but straightforward process. As people began intentionally cultivating teosinte, instead of simply gathering seeds from wild plants, they discovered and saved plants with large seeds, plants with greater numbers of seed rows on a cob, and plants that retained their seeds on the cob instead of dispersing them naturally. They also found plants that matured at the right time, were tolerant of drought, had acceptable flavor and texture, and resisted local insect pests and diseases. The end result was domesticated corn with ears much smaller than those we now have, and with colored kernels.

After domesticated corn developed in central Mexico, people began carrying it both northward and southward from its center of origin. As they did, over many generations, they practiced mass selection where they lived, selecting types that were well adapted to the local climates. By 1000 CE, when the Ancient Pueblo People were cultivating corn on the same land where my grandparents later did the same, they had varieties of corn that were tolerant of the frequent aridity and heat of that area. Unfortunately, the ancient corn

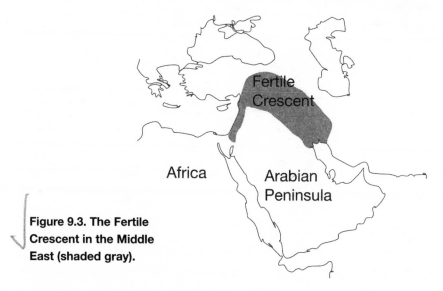

Figure 9.3. The Fertile Crescent in the Middle East (shaded gray).

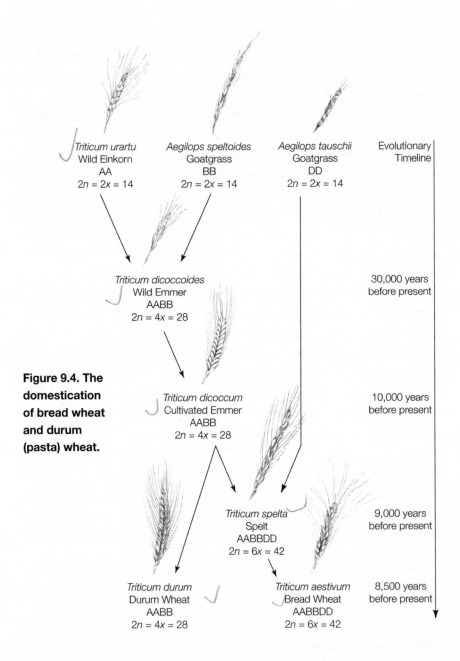

Triticum urartu
Wild Einkorn
AA
$2n = 2x = 14$

Aegilops speltoides
Goatgrass
BB
$2n = 2x = 14$

Aegilops tauschii
Goatgrass
DD
$2n = 2x = 14$

Evolutionary
Timeline

Triticum dicoccoides
Wild Emmer
AABB
$2n = 4x = 28$

30,000 years
before present

Figure 9.4. The domestication of bread wheat and durum (pasta) wheat.

Triticum dicoccum
Cultivated Emmer
AABB
$2n = 4x = 28$

10,000 years
before present

Triticum spelta
Spelt
AABBDD
$2n = 6x = 42$

9,000 years
before present

Triticum durum
Durum Wheat
AABB
$2n = 4x = 28$

Triticum aestivum
Bread Wheat
AABBDD
$2n = 6x = 42$

8,500 years
before present

seeds found among their ruins have long since lost their viability—none can be germinated to reproduce the corn that existed then.

This process of plant and animal domestication took place at roughly the same time throughout the world, in some cases with considerable complexity. Wheat, for example, was domesticated in the Middle East in what is known as the Fertile Crescent, an area located in parts of modern-day Iraq, Turkey, Syria, Israel, Lebanon, Jordan, Iran, and Kuwait (figure 9.3, page 227). In this region, wild grasses related to wheat are common, and three of them were the parental species of modern wheat.

Figure 9.4 (page 228) illustrates what happened as an early form of wheat arose in nature, which humans then domesticated. About thirty thousand years ago, two wild grass species, wild einkorn and a type of goatgrass, hybridized naturally in many places. The hybrids were infertile for the same reasons that mules are infertile—chromosome arrangements that differ between the two species. However, in plants it is not uncommon for chromosome numbers to double, which restores fertility to the hybrids. In the case of wheat, the chromosome-doubled hybrid is called wild emmer, and hybridization followed by chromosome-doubling probably happened many times in nature to produce multiple populations of wild emmer.

Ancient people began cultivating wild emmer, practicing mass selection for traits that improved the emmer about ten thousand years ago in the Fertile Crescent. The result was cultivated emmer, which a few people in parts of the Middle East and Europe still grow on small farms.

As people began raising cultivated emmer, they carried it northward, expanding its range into modern-day Turkey. By about nine thousand years ago, it had entered the range of another related wild goatgrass species. This wild species hybridized with cultivated emmer, producing yet another infertile hybrid that regained fertility when its chromosomes doubled. The resulting chromosome-doubled hybrid had six sets of chromosomes, two from wild emmer and two from each of the two species of goatgrass. This domesticated species is known as spelt, which is also still grown as a specialty crop for making spelt bread.

The seeds of both cultivated emmer and cultivated spelt are encapsulated in a hull, which strongly adheres to the seed. However, about 8,500 years ago, a mutated type of spelt appeared that allowed the hulls to easily separate from the seeds when people threshed the plants. This was a major development because it enabled people to grind the hull-free grain to make flour. Through mass selection, ancient farmers in the Middle East developed this type of free-threshing wheat into bread wheat, the same type of wheat we now use to make bread.

In the meantime, the original cultivated emmer was still being grown in the southern part of the Fertile Crescent, but it was not a free-threshing type. Then, about seven thousand years ago, a mutation produced a free-threshing type of cultivated emmer, much like the mutant type in spelt. Through mass selection, ancient farmers developed this new type of emmer into a wheat now known as durum wheat. This wheat has four sets of chromosomes derived originally from wild emmer and one goatgrass species and is a different species than bread wheat. Today, durum wheat is the type used for pasta products and couscous. The famous semolina wheat in Italian pasta is a type of durum wheat.

When scientists studied in detail the different species of cultivated wheat and their wild relatives, they found evidence of several chromosome-shuffling events, very similar to those in human and great ape chromosomes discussed in chapter 6. For example, modern bread wheat has two chromosome translocations and two inversions when its chromosomes are compared to its wild relatives, one of several characteristics that evolutionarily distinguish it from its parental species.

Continuing on this same trend, scientists have been able to reconstruct the chromosomal shuffling that took place among the several domesticated grass species—among them wheat, rye, barley, rice, maize, millet, sorghum, and sugarcane—showing how ancient people dramatically shaped the evolution of the plants they used for food. And now, we are the beneficiaries.

DOMESTICATION TOOK PLACE IN DEFINED CENTERS OF CIVILIZATION.

Although people had spread to many parts of the world by ten thousand years ago, domestication of plants and animals was, for the most part, localized to a few regional centers. Not coincidentally, these are also the centers of major ancient human civilizations. The Russian geneticist Nikolai Vavilov made more than a hundred expeditions during the 1920s and '30s throughout the world, documenting the domesticated species and their wild relatives from these centers. Although Vavilov focused his research on plants, most domesticated animal species originated in the same centers as those he identified for plants.

Vavilov identified eight centers (figure 9.5). Center I is in East Asia in what is modern-day China. Among the species domesticated there are soybean, peach, lime, and the yak. Center II is in south and Southeast Asia and is the center where rice, banana, orange, lemon, tangerine, cucumber, sugarcane, chicken, and the water buffalo were domesticated. Center III is in south Asia, mostly in modern-day India and Pakistan, and is the center of domestication for apple, apricot, cherry, onion, and pear. Center IV, the Fertile Crescent in the Middle East, is a major center for domestication of a wide variety of species, including wheat, barley, oats, rye, fig, grape, lentil, pea, spinach, Old World cotton, cattle, goat, sheep, pig, and horse. Center V is the region near the shoreline of the Mediterranean Sea, which also has given us a wide range of domesticated species, among them artichoke, beet, broccoli, Brussels sprout, cabbage, carrot, celery, chard, flax, lettuce, olive, radish, camel, donkey, and rabbit. Center VI, in sub-Saharan Africa, is the place of origin of millet, sorghum, cantaloupe, watermelon, yam, coffee, and sesame. The remaining two centers are in the Americas. Center VII, in Middle America in the area of modern-day Mexico and Guatemala, is where corn, some species of beans

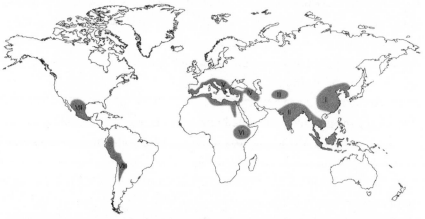

Figure 9.5. The eight centers of domestication identified by Nikolai Vavilov.

and tomato, avocado, cassava, papaya, pecan, peppers, pumpkin, most species of squash, sweet potato, and the domestic turkey originated. Center VIII is in the Andes Mountains of South America in what is modern-day Bolivia and Peru. Potato, some species of beans and tomato, New World cotton, peanut, llama, and alpaca were domesticated there.

The coincidence of these regions with major ancient civilizations is readily apparent. The great Chinese dynasties arose in Center I, and Center II was the site of several civilizations, among them the Ayutthaya Kingdom, the Khmer Empire, the Srivijaya, and the Majapahit. The ancient civilizations of India, among them the Indus Valley and Vedic civilizations and the Mahajanapadas, arose in Centers II and III. The Assyrian and Babylonian Empires ruled region IV, and the Roman, Greek, Hebrew, and Egyptian civilizations thrived in Center V. The Kush civilization in what is now Sudan, and the Egyptian Empire ruled much of the region in Center VI, which is also the most probable place of origin for modern humans. The Aztec and Maya civilizations arose in Center VII, and the Aymara and Inca Empires in Center VIII.

Did these civilizations produce domesticated species, or was it the other way around? In cases such as this, the answer is often not a simple either-or situation but rather a complex interaction of both. Civilization and domestication of plants and animals are interdependent; one builds upon the other. However, the results from archaeological studies are quite clear—agrarian culture preceded domestication, which preceded the emergence of major civilizations. Once civilizations were established, farmers continued to further domesticate the plants and animals they cultivated. We are still doing so today with modern plant and animal breeding.

Although the plants and animals we use for food originated in specific regions, many have since spread well beyond their origins and have become central to certain cultures. For example, it is hard to imagine Italian food without tomatoes (which originated in the Americas). I recall reading a shopping list written by Michelangelo, and conspicuously absent from it was anything having to do with tomatoes. Only well after European explorers returned from the Americas with tomatoes in the sixteenth century did Italians begin to adopt them as part of their cuisine.

When people began traveling from one continent to another, carrying different plants and animals with them, they introduced an entirely new evolutionary scenario, exposing these plants and animals to pests and pathogens they had never before encountered. The potato is an example—and the consequence was one of the greatest tragedies in recent human history.

Potatoes were domesticated in the Andes Mountains of South America. Spanish explorers brought them to Europe, where they soon became an essential staple food for the rural poor, as is evident in Vincent van Gogh's painting *The Potato Eaters*.

By the mid-nineteenth century, potatoes had become a major source of food in Ireland. Meanwhile, across the Atlantic Ocean in Mexico, a fungus naturally infected a wild Mexican native plant that is distantly related to potatoes. The wild plant and the fungus had been coevolving for thousands of years with resistance in the plant countered by the fungus overcoming the resistance. Potatoes, which were native to South America, had never encountered this fungus because it had not been transported to South America. However, fungal spores hitchhiked to Europe with Spanish explorers, where, for the first time, the fungus encountered the potato plant. Because potatoes had not previously been exposed to this fungus, natural selection had never favored

Figure 9.6. *The Potato Eaters* by Vincent van Gogh.

resistance to the fungus in potatoes. When the two finally met in Europe, potato plants readily succumbed to fungal infection—a disease named *late blight* because it strikes potato plants late in the growing season, causing them to rot in the fields. During the years 1845–52, late blight spread throughout potato fields in Ireland, resulting in the Great Irish Potato Famine. Millions of people starved—some succumbing to death, others emigrating elsewhere in the world to escape the famine.

Another instance of evolution is now under way in North America as a consequence of humans carrying food plants from one continent to another. The apple was domesticated in India and Pakistan (Vavilov's Center III). It made its way to Europe and eventually to North America, where a new insect pest soon began to infest it. A natural pest of native American hawthorns is a fly, called the hawthorn maggot, which lays its eggs on the fruits of the hawthorn bush, a distant relative of apples. The eggs hatch into maggots, tiny worms that bore into the fruit and consume it. In the mid-nineteenth century, some of these flies laid eggs on domesticated apples growing in the Hudson River valley of New York. These eggs hatched into maggots that fed on the apples instead of on hawthorn fruits. Now, more than a century and a half later, the apple maggot has evolved from its ancestor, the hawthorn maggot, to thrive exclusively on apples. It is a new apple-specific insect pest well on its way to becoming a new species.[2]

DOMESTICATION HAS GIVEN US A WEALTH OF FOOD SPECIES WAITING TO BE REDISCOVERED.

A large number of domesticated food species—in fact, a majority—have not become worldwide successes. I have spent much of my career researching a domesticated plant called quinoa, which was a major food staple for the Inca and Aymara civilizations of Andean South America (Vavilov's Center VIII). It originated from wild weedy plants common throughout the Americas. For several reasons, quinoa never took hold in Europe, unlike the potato, which was domesticated in the same region as quinoa and gained wide European acceptance. The Spanish displaced quinoa cultivation in the Andes when they enslaved the native people and forced them to plant wheat and barley, instead of quinoa, for making bread and brewing beer. Also, most varieties of quinoa are adapted to high-altitude conditions of the Andes Mountains near the Tropic of Capricorn and do not grow well in the low elevations and high latitudes of European farmland. It is unfortunate for the rest of the world that quinoa did not spread beyond South America because it is the most nutritious food known, and it is highly resistant to drought, saline soils, and frost. Had it become adopted worldwide as a major food, much hunger and malnutrition suffered in today's world might have been prevented.

My own research with quinoa has focused on collaborating with Bolivian and Peruvian scientists to genetically improve it so that it can better feed impoverished people in its region of origin. To do so, we rely heavily on our understanding of its evolutionary history.[3] The increased production resulting from this research has helped impoverished Bolivian and Peruvian farmers to earn a modest cash income by exporting quinoa for an increasingly popular market. With some scientific development, people elsewhere in the world may someday be able to cultivate it, especially in arid and saline regions where other crops typically fail. Before this happens, there are many cultural, political, economic, and biological barriers to cross.

Ancient humans domesticated hundreds of other lesser-known species of plants and animals, some of which have gained worldwide popularity only within the last century. The United States and Brazil, for example, are major soybean producers, and soybean products—vegetable oil, protein, carrageenan, lecithin, and others—are ubiquitous in processed foods. Yet, in the first half of the twentieth century, soybeans were cultivated mostly in the Far East for products consumed there. In the 1950s and '60s, soybean production spread throughout much of the world as soy products gained a foothold in the food-processing industry.

Before the 1970s, few people were aware of kiwifruit, which is now common on grocery store shelves worldwide. Many more wonderful foods—the products of thousands of years of domestication by our ancient ancestors—are treasures awaiting rediscovery.

GENETIC EROSION IS A THREAT
TO OUR FOOD SUPPLY.

In northern Moravia near the Czech-Polish border is a quiet village called Hynčice. Here a young boy named Johann Mendel learned from his parents how to cultivate crops and fruit trees. The local schoolmaster recognized the boy's innate brilliance and convinced his parents to send him to boarding school. As Johann grew older, one of his teachers arranged for him to enter a Roman Catholic monastery in the city of Brno. Upon joining the Augustinian order there, he changed his name to Gregor Mendel and began teaching natural science at a nearby school.

Mendel took leave from the monastery to study at the University of Vienna, where he was fortunate enough to work under some of the most prominent scientists of his day, among them the botanist Franz Unger and the physicist Christian Doppler. Upon his return to the monastery two years later, Mendel initiated a series of experiments with peas and beans that would eventually revolutionize our understanding of heredity and evolution.

Mendel himself was a well-versed scholar of plant breeding. Books in the monastery library on plant breeding contain his notes, and their influence on Mendel's work is immediately apparent.[4] Plant breeders in Mendel's day had taken their science well beyond mass selection. They carefully chose desirable plants and intentionally hybridized them, selecting in subsequent generations the best types that appeared. Mendel discovered the underlying biological reasons for success in plant breeding, and his discoveries eventually established (after his death) the modern sciences of plant and animal breeding.

Applying Mendelian principles, plant breeders made rapid and dramatic improvements in the plants we use for food beginning in the early twentieth century. Animal breeders likewise applied Mendelian principles to rapidly improve the productivity and vigor of our domesticated animals. Today's domesticated plants and animals are far more productive than they were a century ago, largely because breeders have substantially changed their genetic constitution through modern breeding methods.

In 1798, in his *Essay on the Principle of Population*, Thomas Malthus predicted a dire consequence for humanity. He surmised that because human populations were growing exponentially, whereas food production was increasing linearly, at some point human population growth would so greatly surpass food production that mass starvation would be the result. Darwin read Malthus and referred to him in his writings. As it turned out, Malthus was half right. The world's human population *has* continued to grow exponentially, from about 980 million during the time when Malthus lived to more than seven billion today.[5] Malthus, however, did not anticipate the effect of modern breeding and agricultural methods on food production. As human populations grew exponentially, so did food production. In fact, agricultural science, breeding in particular, has been so successful that increases in food production have *exceeded* increases in population growth to the point that we now produce more food than at any time in the history of the world. But the current world human population is so large that our ability to produce enough food may not be sustainable.

Moreover, about one-fifth of the world's population suffers from severe hunger and malnutrition. Here and now, there is enough food in the world to feed everyone, so some people argue that the problem is not food production but distribution. Unfortunately, such an argument is overly simplistic and misinformed. In most places where people suffer from severe hunger and malnutrition, food is readily available. But those who suffer cannot afford it. The true cause of hunger, in the vast majority of cases, is poverty. Most people who suffer from severe hunger are themselves farmers, typically people who have small tracts of land and little access to the technologies that have revolutionized food production in modern times. Through modest investments in appropriate technologies and micro-credit, impoverished farmers can produce more food and achieve a modest income, and by so doing can begin to escape hunger and poverty. What we most need to overcome world hunger is the political will to appropriately make those investments.

The development of highly productive plant varieties and animal breeds through modern breeding is largely responsible for the dramatic increases in food production during the twentieth century. Few people are aware that modern breeding also has a serious downside, one that is directly related to evolution, and one that continually and increasingly threatens the world's food supply. Plant and animal breeders have made enormous improvements in the genetic productivity of our domesticated species by rapidly identifying and selecting those genetic combinations that are most productive. By doing this, they have discarded much of the inherent genetic diversity in these species, favoring those few genetic variations that offered the highest production.

Most of this breeding took place in North America and Europe, for the most part outside the centers of origin of domesticated species. For example, North American plant breeders developed new, highly productive varieties of wheat, corn, potatoes, and other foods in the United States and Canada, while the original genetic diversity for these species remained in the centers of origin.

For example, in the United States, we typically find only a few types of potatoes in our markets—highly productive varieties of red, white, yellow, and russet potatoes. By contrast, in Bolivia, Peru, and Ecuador, where the potato was domesticated, farmers grow potatoes that make our American potatoes look drab by comparison. South American potatoes have an almost infinite variety of colors, shapes, and sizes—red, orange, yellow, brown, tan, purple, blue, white, in multiple shades, some of them mottled. Potatoes are large and small, some spherical, others long and spindly, some with branches or lobes.

In the 1920s, Nikolai Vavilov, whom we encountered earlier in this chapter, recognized that modern plant breeding had created a serious dilemma, one that threatened its very success. The highly productive plant varieties developed in North America and Europe were making their way back into their centers of origin where they threatened to displace the inherent genetic diversity in the crops grown in those centers. For instance, when Bolivian or Peruvian farmers adopt newer potato varieties developed in the United States, they often reap more abundant harvests but no longer grow the diverse varieties that their ancestors cultivated for generations. Once those older varieties are lost, the genes they carry are also lost forever to humanity. The same happens when new animal breeds are introduced into their centers of origin— genetic diversity is lost forever.

This process is called *genetic erosion*—the irreplaceable loss of genetic diversity caused by displacement of older genetically diverse varieties of plants or breeds of animals with modern genetically uniform ones. It is an alarming irony that the dramatic success of modern plant and animal breeding can potentially eliminate the genetic diversity so essential for its continued success.

HOW CAN WE PREVENT GENETIC EROSION?

To answer this question, we return to Vavilov, who recognized in the 1920s that action was needed to prevent genetic erosion in our food crops. He also realized that there was no reasonable way to prevent modern plant varieties from reaching farmers in centers of origin. To deny these farmers productive varieties was to deny them the means of escaping poverty. Instead, Vavilov took the approach that anthropologists and museum curators take. To see how, let's return for a moment to my grandparents' hometown, where a large Native American burial site has rested untouched for centuries. As settlers arrived and the modern town grew, amateur collectors unknowingly threatened the integrity of the site through their desire to collect, sell, save, and distribute artifacts. To preserve the valuable ancient artifacts at this burial site, professional anthropologists carefully excavated the site and removed the artifacts to museums where they could be cataloged, preserved, and studied. Like these anthropologists, Vavilov and his coworkers made more than a hundred expeditions to centers of origin; collecting seeds, tubers, roots, and cuttings of the diverse crop varieties present in these centers. They carefully cataloged them, storing them in a central facility, much like a museum, in what was then Leningrad (now Saint Petersburg).

Tragically, Vavilov's work was cut short when he became a victim of Stalin's purge of intellectuals just before World War II. More than two thousand artists, writers, and scientists were victims of Stalin's police, who arrested people suspected as threats to the regime. Those who had been educated in Western Europe, Britain, or the United States and those who were perceived as promoting Western intellectual traditions became targets. Vavilov was one of them. In 1940 he was arrested and imprisoned.

The Nazi siege of Leningrad during the winter of 1941–42 was one of the most horrific events of World War II. With all supply lines cut off by Nazi forces, more than one hundred thousand innocent civilians died of starvation when food supplies ran out. Within the institute where Vavilov's collections were housed were seeds of wheat, corn, rice, peanuts, and many other crops, along with potato tubers and other plants that could be eaten. They were too few to make any significant impact on the widespread hunger in the city, but there was enough to keep the workers in the institute from starving—if they ate them. However, eating these stored seeds would destroy the collections Vavilov and his colleagues had assembled over dozens of years, and the genetic diversity held within those seeds would be lost forever. The workers made a pact, agreeing that they would protect the seeds for the future of humanity, sacrificing their lives to do so if necessary. Vadim Lekhnovich, one of the scientists at the institute who survived the siege, described their horrific plight:

> It was hard to walk. . . . It was unbearably hard to get up every morning, to move your hands and feet. . . . But it was not in the least difficult to refrain from eating up the collection. For it was *impossible* [to think of] eating it up. For what was involved was the cause of your life, the cause of your comrades' lives.[6]

Nine of the workers died of starvation, and others from sickness or shrapnel, while protecting the collections. Gary Nabhan recounts this feat in his book, *Where Our Food Comes From*:

> Among this group, Alexander Stchukin died at his writing table, holding in his hand a packet of his most prized peanuts that he had hoped to send off for a grow-out. The custodian of Vavilov's many oat collections, Liliya Rodina, died of starvation, as did Dimitry Ivanov, who as his own life failed, stowed away thousands of packets of rice that he had held so dear. There were others as well—Steheglov, Kovalevsky, Leonjevsky, Malygina, Korzun—some who perished by starving, some riddled by sickness, others by shrapnel. Wolf, the herbarium curator, was hit by a missile shell fragment, and bled to death. Gleiber, the archivist of Vavilov's field notes, died in the midst of those papers rather than leave his post vulnerable to the infidels.[7]

Apart from his colleagues, trapped in Saratov Prison, Vavilov died of malnutrition in 1943. And he was not alone. Dozens of geneticists who had likewise been arrested and imprisoned were executed or died in prison. A generation later, their students made pilgrimages to Mendel's garden, where they collected handfuls of soil. Then, returning home, they scattered the soil on their teachers' graves in tribute.[8]

The sacrifice of these heroes was not in vain—the institute and its collections remained intact and became a model for the rest of the world. Today, the genetic diversity of our food plants and animals is preserved in institutions like Vavilov's, which are often called *gene banks*. Strategically located throughout the world, they serve as resources for plant and animal breeders who rely on the diversity held in them to genetically protect and improve the species we use for food.

WHY IS GENETIC DIVERSITY SO IMPORTANT?

Once a new species has evolved, genetic diversity tends to accumulate in it whether it is wild or domesticated. For wild species, this diversity serves as a buffer to preserve the species. If a new type of disease evolves, a high degree of genetic diversity increases the chances that some individuals of the wild plant or animal species will carry inherited resistance to the disease. The resistant individuals preferentially survive and reproduce through natural selection, and they pass the genes conferring resistance to their offspring. Eventually, the resistance spreads throughout the species as natural selection favors resistant individuals, and the disease is kept in check.

The same is true for domesticated species. They, too, are subject to devastation by diseases and pests, and when the devastation is widespread, food shortages and famine may result. Although history is replete with crop disease and pest pandemics resulting in famine—the Irish potato famine is one of many examples—rarely have such outbreaks occurred in recent times. The reason has to do with modern plant and animal breeding. Scientists regularly monitor the evolution of new crop diseases throughout much of the world. As soon as a new strain of a disease appears, breeders begin searching for genetic resistance in the plants they breed. To find it, they often must sort through the genetic diversity in gene banks. Were it not for this diversity, disease pandemics, crop failure, and food shortages could be the result.

As an example, let's turn to stem rust in wheat, one of the most dev-astating, famine-causing crop diseases in history. When stem rust hits, a moldy fungus infects the stems of growing wheat plants. Its brownish-orange color makes the stems look like they are rusting away, hence the name. The stems collapse, and a field of once-healthy green wheat can become a tangled brownish-orange mess, with no harvestable grain. Winds pick up the dusty fungal spores, carrying a rust-colored cloud to wheat fields both near and far, spreading the disease, sometimes across entire continents.

Throughout much of human history, stem-rust pandemics have periodi-cally destroyed wheat crops. In the twentieth century, scientists discovered genes that conferred resistance to stem rust in the diverse types of wheat stored in gene banks. These resistance genes, however, were specific to par-ticular races of the fungus. As plant breeders developed new rust-resistant wheat varieties, the fungus soon evolved into new genetic races that overcame the resistance. Wheat breeders were forced to search through gene banks to find yet more genes conferring resistance. For decades, they played a cat-and-mouse game with stem rust, repeatedly searching for and finding new resis-tance genes as new fungal races evolved. One of the most serious stem-rust pandemics struck North America in 1953 and 1954. It began in Kansas, then spread northward across the Great Plains and into Canada. In those two years, much of the previously resistant wheat crop had succumbed to rust, resulting in a loss of about five million tons of wheat.[9]

About that time, the scientists who were persistently searching for new resistance genes discovered something new. They found in gene banks an entirely different type of genetic resistance—genes that conferred what is known as durable resistance, which resists a broad range of fungal races, and lasts. The stem-rust fungus was unable to overcome durable resistance.

This discovery came at a time when the Rockefeller Foundation had sent US scientists to Mexico to help that country overcome its serious problems with poverty and hunger. Their charge was to bring modern methods of agri-culture to increase food production. Stem rust had repeatedly plagued Mexico since the 1500s when the Spanish introduced wheat into that part of the world. Unlike more advanced countries, Mexico did not have a sustainable scientific program for breeding rust-resistant varieties of wheat. The wheat varieties grown there in the mid-twentieth century were mostly descendants of the old rust-susceptible Spanish types introduced centuries earlier.

Norman Borlaug, one of the Rockefeller scientists, led a research team focused on combating stem rust in wheat. When the new class of durable-resistance genes became available, he and his team bred new varieties that carried these genes. They also introduced dwarfing genes into the new varieties to make them short-statured and strong, able to resist collapsing in the wind. Moreover, they bred the new varieties to succeed in a wide range of environments. To complement the scientific work, the Mexican government invested in farm infrastructure to modernize farming methods and marketing.

Mexican farmers rapidly adopted the new varieties and modernized their farming. Wheat production soared, turning Mexico from a wheat importer to a wheat exporter. Borlaug and his team took the new rust-resistant wheat varieties to India and Pakistan, where they replicated the Mexican experience, transforming the economies of these countries. Some of the scientists who had worked on the wheat research turned their attention toward rice and likewise boosted rice production throughout much of Asia. The result was a revolution—soon dubbed the Green Revolution—for which Borlaug received the Nobel Peace Prize in 1970.

Meanwhile, in more advanced countries such as the United States, scientists had introduced durable rust-resistance genes into wheat, overcoming the immediate danger of pandemics. The last stem-rust pandemic in the United States was the 1953–54 devastation. Since that time, durable genetic resistance has held stem rust in check—that is, until 1998, when evolution once again took over.

That year a new race of stem rust evolved to defeat durable resistance. It was first detected in Uganda and has since spread to Ethiopia and Kenya in Africa. Then it jumped the Red Sea and spread into the Arabian Peninsula. Unless it can be stopped, it will eventually spread throughout the world, threatening the world's entire wheat crop. More than 90 percent of all modern wheat varieties are susceptible to it. Through a massive testing effort, scientists have found a few genes from wheat gene banks that confer resistance.[10] A new mutation in the fungus, however, has already defeated one of them.[11] Fortunately, new resistant varieties are now available, but it's uncertain whether their seed can be multiplied fast enough to halt the disease's onslaught.[12]

PEST EVOLUTION ALSO THREATENS OUR FOOD.

As new species of plants and animals evolved when people domesticated them, new species of pests that feed on these domesticated species coevolved with them. As a consequence, we now have hundreds of pests named after the plants and animals they infest—Colorado potato beetle, corn earworm, rice seed midge, soybean cyst nematode, pea weevil, cattle grub, chicken mite, and hog sucking louse are just a few examples. Ancient people probably grouped their plants and animals in proximity in garden plots, farm fields, cages, or corrals, providing an environment in which pests could proliferate by easily jumping from one plant or animal to another. Modern agriculture, with enormous fields stretching for miles on end and gigantic feed lots or buildings with chicken cages with hundreds, even thousands, of animals in proximity, has created havens for proliferation of insect pests and a need for us to control them.

Insect plagues have afflicted the crops and animals people raise since ancient times. For instance, Roman, Greek, Assyrian, Babylonian, Egyptian, and Hebrew literature is replete with accounts of locust plagues. Clouds of large grasshoppers sweep into a wheat field, and within a few hours the insects eat the plants to the ground then move on to find other fields to decimate. People struggling to save their crops in a futile effort to beat the insects to death or sweep them away are staples of ancient legend and some modern news reports.

At yet another level of coevolution, pests of pests have evolved. As an example, let's look at aphids, tiny sap-sucking insect pests that infest crop, vegetable, and ornamental plants. They prolifically reproduce through parthenogenesis, a process by which females almost incessantly give live birth to females genetically identical to themselves by naturally cloning themselves without any sexual reproduction. A few aphids arriving in a crop field can reproduce into millions within a few days. Because aphids reproduce so quickly and there are so many of them, they serve as an abundant food source for predatory insects that have coevolved with them—beetles, flies, and wasps are the most common aphid eaters. Perhaps the most famous are ladybird beetles. The attractive black-spotted red, orange, or yellow ladybugs so familiar to children (and adults) who have no hesitation allowing them to crawl on their hands, arms, and clothing, are ravenous aphid eaters, each of them capable of consuming as many as fifty aphids per day. For all of human history, even before our ancestors domesticated plants and animals, pests of pests have evolved. They protected the food of our ancestors and still do so today.

Modern mechanized agriculture, with its large fields and feed lots, relies to a large extent on pesticides, usually insect-killing chemicals. Although most people tend to think of pesticides as modern inventions, their use dates back thousands of years. Sulfur, lead, arsenic, and mercury were used in ancient times to kill insects on food plants and animals—no doubt killing some people and animals as well. By the seventeenth century, scientists discovered that extracts from some plants, such as tobacco and chrysanthemum, were effective as pesticides. Commercial growers and home gardeners still use nicotine (from tobacco) and pyrethrum (from chrysanthemum) to control insect pests.

One of the most widely used pesticides during the mid-1900s was DDT. It controlled nearly every type of insect pest and was widely used in agriculture. It also was effective for mosquito control, particularly as a way of reducing malaria in tropical regions. During the 1940s, it combated and even eliminated certain insect-borne diseases, such as typhus and malaria, in many parts of the world. For example, it contributed to the allied victory in World War II as specialists in the armed forces sprayed it throughout insect-infested areas to protect troops from typhus in Europe and malaria in the South Pacific.

As a pesticide, DDT was thought to be completely safe—black-and-white newsreels from the 1940s depict children in swimsuits happily dancing under a shower of DDT spray from mosquito-control tanker trucks patrolling the streets of US cities. However, by the late 1950s, scientific studies revealing the adverse health effects of DDT on humans and birds began to emerge. Rachel Carson published her landmark book *Silent Spring* in 1962, chronicling the threat of DDT to humans and the environment. The book's popularity instigated yet further scientific investigation, which confirmed the health and environmental threats of DDT and other pesticides. In response to the ensuing public outcry, the United States banned DDT in 1972.

By then, however, DDT had lost much of its effectiveness against insects, thanks to the predictable process of evolution through natural selection. When applied over large areas, nearly all insects exposed to DDT died. A few, however, were naturally resistant to it, and they survived. The resistance was inherited, conferred by genes transmitted to offspring. As resistant females and males mated with each other under exposure to DDT, natural selection favored the most resistant offspring. Populations of increasingly DDT-resistant insects evolved, prompting farmers and pest-control agencies to apply even higher doses of the insecticide. Through natural selection, insects with yet stronger resistance emerged. The same has happened for insects exposed to other widely used insecticides, as well as for rats and mice exposed to rodent poisons.

THE STORY OF *BT* IS AN ONGOING SAGA OF EVOLUTION.

Like all animals, insects suffer from disease-causing infectious microorganisms. One of the most widespread is the bacterial species *Bacillus thuringiensis*, more commonly known as *Bt* (pronounced "bee-tee"). The bacteria have a broad number of hosts—they infect and kill the larvae of moths, butterflies, and many beetle species. The bacteria are harmless to humans. Generations of people have been naturally exposed to *Bt* long before chemical pesticides were known. Because it is a natural organism, instead of a synthetic chemical, *Bt* is officially approved and often used for pest control in organic food production. It is available in many garden stores for home gardeners, and commercial organic-food growers routinely apply it to the plants they raise.

Some of the most serious insect pests are infected by *Bt*. As beautiful as adult butterflies and moths are, their larvae—multi-legged caterpillars—are voracious leaf, fruit, and seed eaters. The worms that bore into apples or consume the tender kernels at the tip of a corn cob are often moth larvae. Beetle larvae are also destructive. Weevils are beetle larvae capable of destroying plant products as diverse as kernels of wheat and cotton bolls. But they are susceptible to *Bt*, a safe and natural pesticide.

Bt produces a toxin that kills butterfly, moth, and beetle larvae. The toxin is a protein, which is encoded by a single gene—a condition that some innovative scientists recognized as being ideal for genetic engineering. Because of the ancient, common evolutionary origin for all of life, the genetic code is the same throughout all of life, so a gene belonging to one species encodes the same protein if that gene is transferred to another species—even species as disparate as a bacterium and a plant. If scientists could put the *Bt*-toxin gene into crop plants, these plants would be naturally immune to any insect susceptible to *Bt*.

Moreover, doing so could be an enormous benefit to the environment and human health. Farmers often spray chemical pesticides to control moth larvae in cotton, corn, soybeans, and many other crops. Traditionally, cotton has been one of the most pesticide-treated crops because it is a nonfood crop. During my high school years, I lived in southeastern Arizona. Our house was situated on about two acres of land where we had a garden, and I raised chickens, quail, and a goat. Adjacent to our land was a cotton field. On summer mornings, I often awoke to what sounded like a World War II bombing run as crop-duster airplanes flew just above the tops of the cotton plants, spraying pesticides on them. At the end of the field, pilots pulled their planes skyward, flying so close over our house that I could see them wave at me through the plane's window when I stood in our yard. Pesticide spray drifted from the plane onto our house and yard, its chemical smell giving me headaches.

The pesticides used in those planes were organophosphates, potent neurotoxins that attack the nerves of both insects and people. I still worry about long-term effects of the pesticide exposure I suffered during those years. If someone could make the cotton immune to its pests, the spraying could cease, and we'd all be better off for it.

That's what happened in 1996—the first *Bt* cotton was released. Genetic engineers had extracted the *Bt*-toxin gene from bacteria, reengineered it to function in cotton, and inserted it into the DNA of cotton plants. These genetically engineered plants produce the same *Bt* toxin as the bacteria in their leaves, flowers, stems, roots, and—most importantly—in the bolls, the rounded fruits containing the growing cotton fibers. Before 1996, tobacco budworm, cotton bollworm, and pink bollworm were serious cotton pests. In 1995, the year before *Bt* cotton became available, 29 percent of Alabama's cotton crop was lost to these pests, even though Alabama cotton farmers applied the highest amount of chemical pesticide in the United States.[13] When these larvae begin chewing on a cotton bud or boll on a *Bt* cotton plant, they immediately die from the toxin before they can damage the cotton, and no chemical pesticide is needed to control them—the gene that encodes the organic pesticide is within the DNA of every plant in the field.

These *Bt* cotton plants are genetically modified organisms, more popularly known as GMOs. Technically *all* domesticated plants and animals are genetically modified organisms because they all have been subjected to intentional breeding since they were first domesticated thousands of years ago and especially in the past century through modern plant and animal breeding. However, the term *GMO* applies only to organisms that contain nonnative genes introduced through genetic engineering in a laboratory. Because *Bt* cotton has an artificially inserted bacterial gene, it is a GMO. The cotton grown in the fields when I was young was not a GMO. It is called Pima cotton, a type subjected to intense modern breeding and derived from types originally grown by ancient Native Americans in the same area.

The proponents of GMOs argue that not only does *Bt* cotton protect people; it also protects the environment, and with good reason. Chemical pesticides are indiscriminate; they kill nearly all insects—only those few insects resistant to the pesticide escape. Harmful and beneficial insects alike die when exposed. For example, my father, wanting to raise our own honey to sell, bought some active beehives in white wooden boxes and placed them behind our chicken coop. For a short while we harvested fresh honey. But when the planes began their runs in the cotton fields, our bees died and our family's brief foray into honey production failed. The bees were visiting cotton flowers to collect nectar and pollen. When the planes arrived, the deadly insecticide indiscriminately killed the harmful budworms and bollworms—and the beneficial honeybees.

Had *Bt* cotton been available then, our bees might have survived. *Bt* cotton kills only the insects that feed on the plants—all other insects are unaffected. However, our bees were feeding on the cotton plants, collecting nectar and pollen from the flowers. Had those plants been *Bt* cotton, the nectar and pollen might have carried *Bt* toxin. But the *Bt* toxin targets larvae of butterflies, moths, and beetles, so bees escape the danger.[14]

Even *Bt* spray, used extensively in organic farming, has its environmental downside. When sprayed on plants, it does not discriminate between harmful larvae of pests and harmless butterfly larvae. For example, in the early 1990s, crop-duster airplanes and helicopters loaded with *Bt* spray flew over cities in the United States, applying clouds of *Bt* bacteria on trees to control gypsy moth larvae. Such widespread spraying also killed butterfly larvae. Soon even common butterflies, such as monarchs, swallowtails, and painted ladies, began to disappear from areas where they had been abundant. Of even greater concern were rare and endangered butterflies, which also died from *Bt* exposure. A genetically engineered crop, by contrast, kills only those larvae that actually feed on it—no spraying required.

Bt corn is now widely planted in the United States where laws governing GMO crops are quite lenient. Most North Americans who eat corn products, such as corn chips or tortillas, routinely consume *Bt* corn. No special labeling is required for GMO plant products in the United States, Canada, and many other countries. The European Union, by contrast, requires labeling of foods containing GMO products. The reason is public concern for safety—concerns that GMO foods may be harmful to people. The *Bt* toxin is essentially the same in GMO plants as it is in its natural form in bacteria, and its safety for human consumption has long been known. Thus far, in spite of multiple studies, there is no credible evidence that GMO plants are harmful to humans.

There are, however, legitimate concerns about GMOs. In the case of *Bt* crops, it is not any potential harm to human health that should give us concern but the natural, ongoing process of evolution. Widespread use of *Bt* crops will produce an inevitable outcome—*Bt*-resistant insect pests will evolve, just as they have evolved to resist widely applied insecticides. Such an outcome will not only render *Bt* crops ineffective; it will also be a major blow to organic farmers and gardeners, as one of the few pesticides they can use will lose its efficacy.

Scientists have stepped in with a strategy to delay the emergence of *Bt*-resistant insect pests, devised from our understanding of how evolution works. Natural selection is most effective when selection pressure is widespread. If farmers *exclusively* grew *Bt* crops, resistance would rapidly evolve. To prevent this outcome, scientists recommend that farmers surround each *Bt* crop field with strips of non-*Bt* plants of the same crop. These strips are refuges where insects susceptible to the *Bt* crop can feed on non-*Bt* plants. Their progeny remain susceptible to *Bt*, and the few resistant insects that emerge from the majority of the field planted to the *Bt* crop are likely to mate with the susceptible insects from the refuge. Genes that confer susceptibility persist in the overall insect population, diluting the genes that confer resistance, and the evolution of *Bt* resistance is delayed.

There is evidence that resistance evolves when this strategy is not employed. The first *Bt*-resistant bollworms evolved in *Bt* cotton planted in the southeastern United States, where refuges are not utilized as extensively as elsewhere. There is also evidence that the refuge strategy is working where it is widely employed.[15]

WHY DOES IT MATTER?

As our ancient ancestors evolved into the most intelligent species on the earth, they domesticated once-wild plants and animals, selecting those individuals with the best characteristics as parents for subsequent generations. These same plants and animals now feed us, provide us with materials for our clothing, ornament our homes, and serve as our pets. In just a few thousand years, these ancient people invented an unprecedented form of coevolution—we have become dependent on these species of plants and animals for our food, just as they have become dependent on us for their survival and propagation.

Another type of coevolution threatens our modern food supply. Just as microorganisms that cause human disease have coevolved with us as their hosts, so also have microorganisms that cause diseases in our food plants and our domesticated animals coevolved with them. Likewise, pests of food plants and animals have coevolved with them. As these infectious microorganisms and pests continue to evolve, natural selection allows them to overcome the steps we take to control them.

Few people realize how dependent we are on modern science for our food. Over the past century, nearly all countries have to some degree transitioned from agrarian to industrial to postindustrial societies. The proportion of people who produce food has dwindled. In some places, such as the United States, most farms are large and highly mechanized. On these farms, genetic uniformity is an asset because genetically uniform plants and animals offer predictable products for marketing. A farmer who plants a particular variety of wheat, for example, knows beforehand the baking qualities of its flour, how tall it will grow, approximately how much it will yield, when it will mature, and its resistance to particular diseases.

Unfortunately, genetic uniformity *promotes* rapid evolution of microorganisms that cause diseases in our food crops and domesticated animals. Scientists constantly monitor these emerging diseases while plant and animal breeders persistently work to incorporate genes that confer disease resistance. The source of these genes is the genetic diversity held in gene banks, and we are dependent on this diversity to produce enough food to support a world population of billions. Without modern agricultural science—much of it based on evolutionary theory and observations—massive food shortages would result.

Moreover, the widespread use of pesticides inevitably promotes the evolution of resistant pests. The same is true when fungicides are used to control fungal disease, or herbicides to control weeds. For example, weeds resistant to popular herbicides, such as atrazine and glyphosate (which is often sold under the popular brand name Roundup®) have evolved in recent years.[16]

As dependent as we are on agricultural science, it is under threat. As an example, the success of durable resistance to wheat stem rust led politicians to cut funding for wheat rust research because the problem of wheat rust seemed to be solved. When a new race of stem rust emerged in 1998, the world was ill-equipped and unprepared for it, so scientists had to scramble to find funding to combat it, losing valuable time as the fungus spread to wheat fields in Africa and the Middle East. The losses that have already occurred when wheat fields succumbed to the new race of rust far outweigh the cost of preventive research that could have averted these losses.

And gene banks are also under threat. So few people are aware of them; they are rarely a political priority. They are not businesses and must operate with funding from governments and nonprofit foundations. Without strong political capital, gene banks often become targets for funding cuts, or for diversion of their facilities to other uses. In 1926, Vavilov established the Pavlovsk Experimental Station as a gene bank for fruit trees and berries, and it currently houses the world's largest collection of these species. Because most fruit trees and berry bushes are propagated through cuttings and not seed, they must be preserved in gene-bank orchards as living plants. Although Pavlovsk Experimental Station survived the Nazi onslaught during World War II, the station is now in danger of destruction. But this time invading armies are not the culprit. The trees and bushes containing the world's largest collection of genetic diversity for fruits and berries are under threat from bulldozers to make way for luxury condominiums. Scientists throughout the world are protesting. We can only hope that their voices are widely heard.

As people express concern over perceived health threats of GMO plants, a lesser-known but far more serious threat lurks in the evolution of *Bt*-resistant pests, as we saw with *Bt* cotton and corn. A similar situation threatens herbicide-resistant GMO crops. When farmers plant GMO crops that are resistant to a particular herbicide, they can use that herbicide to kill weeds while the crop grows. Widespread herbicide use, however, also favors the evolution of herbicide-resistant weeds.

Food is so readily available in our modern postindustrial society that few people consider how seriously evolutionary forces affected the food supply in the past and how they continue to do so today. The natural course of evolution in disease-causing microorganisms and insect pests has caused famine, pestilence, and starvation throughout human history. Our understanding of evolution and how it works allows us now to stem the tide of famine and feed our expanding world population. But it is certain that the ever-present threat of natural selection will eventually overcome our successes, meaning that scientists must endlessly continue their work to ensure the security of our food.

CHAPTER 10

EVOLUTION AND OUR ENVIRONMENT

Towering over the tranquil Ligurian Sea in northwestern Italy are the Apuan Alps. Above the town of Carrara, these mountains appear to have permanent snow cascading down their ravines, but the "snow" is an illusion. It is rubble from quarries, which have produced the world's finest sculpture marble for more than two thousand years. Here the Romans extracted stones for the thousands of statues that adorned their vast empire, especially their capital city. Michelangelo traveled here to select the block from which he carved his magnificent *Pietà*. Today, sculptors from throughout the world flock to this place like pilgrims to carve statues in the translucent white stone. I am one of those sculptor-pilgrims; few experiences can match the sublime joy of carving marble statues in Italy.

Millions of years ago, this place was a shallow seabed. Here, corals and shellfish lived and died, their remains settling to the bottom of the sea and accumulating, layer upon layer. Eventually, they were buried deep beneath the earth, where immense geologic pressure crushed and compacted their remains, turning them into limestone. As continents moved a few centimeters each year, they gradually yet powerfully pressed against each another. Geologic forces subducted the limestone deep beneath the earth, where intense heat and pressure metamorphosed it into white, crystalline marble. Further continental movements then lifted the marble, buckling the earth's crust in that region to raise the magnificent Apuan Alps. The marble, once buried deep beneath the earth, was now exposed as steep cliffs on the mountainsides. Most of the remains of shellfish had been pulverized and melted during the stone's formation, although a few marbleized fossil remnants are still intact.[1]

253

In the meantime, life had been evolving in the seas and on the land as continents drifted around the globe, sometimes separating from each other to form new seas and oceans, other times colliding to uplift great mountain ranges. This drift is still under way. The massive Andes Mountains in South America, for instance, continue to rise, as a continental plate under the Pacific Ocean pushes against the western edge of South America. Earthquakes and volcanic eruptions in Chile, Bolivia, Peru, Ecuador, and Colombia are some of the many consequences of these slow yet immensely powerful geologic forces.

Throughout the world, life has evolved for more than three billion years in response to the ongoing upheavals of nature—moving continents, uplifted and eroded mountains, lakes and seas that disappeared leaving vast plains in their wake, advancing and retreating oceans and glaciers, and changing climates. We humans are the newcomers. And we are every bit as much a product of evolution as the life that preceded us, surrounds us now, and will inevitably emerge in the future.

Every individual of every species impacts its environment, each to a very small degree. The collective impact of many individuals, however, can be enormous. About two and a half billion years ago, photosynthetic bacteria evolved in primordial seas, spawning the greatest atmospheric transformation in the earth's long history. Through a series of genetic alterations favored by natural selection, these prehistoric microbes acquired the ability to capture energy from sunlight and used it to convert carbon dioxide and water into sugar, expelling pure oxygen gas (O_2) as a by-product. Each cell added an excruciatingly miniscule amount of oxygen to the atmosphere, hardly enough to cause any impact whatsoever. But the combined effect of countless trillions upon trillions of these cells over hundreds of millions of years forever changed the earth's atmosphere. Never before had oxygen gas been present on the planet in such quantities—nearly 20 percent of the air we now breathe consists of oxygen gas. The planet we now know, and the life it supports, ultimately arose because of this epic event.

Back then, the only organisms on the earth were bacteria. Coupled with photosynthesis was the ability to reverse the process—to extract energy from sugar, converting it to carbon dioxide and water through a process called aerobic respiration. To carry out this process, these bacteria needed oxygen, which by then was abundant in the atmosphere. With oxygen readily available, aerobic respirators (oxygen-using organisms) overtook the earth as its most successful organisms and became the ancestors of multicellular life, including all plants and animals—and us. The photosynthesizers (plants, photosynthetic bacteria, algae, and others) continue to replenish the oxygen that we and all other aerobic respirators extract from our oxygen-rich atmosphere.

These ancient processes of photosynthesis and aerobic respiration evolved more than two billion years ago in bacteria and represent parts of an intricate interaction we share with other forms of life and with our environment. To see how, let's return to Italy and imagine dining in a small Tuscan restaurant overlooking the Ligurian Sea in late fall, with a wonderfully prepared dish of lasagna *al forno* on the table. The lasagna noodles serve as a source of energy, but the process of energy conversion and the interaction of the living with the nonliving began well before the lasagna was baked. A few months prior, during early summer, hydrogen atoms in the sun fused with one another in an intensely hot nuclear reaction to become helium. The process released energy in the form of sunlight, which sped away from the sun in all directions at 186,000 miles per second, a tiny fraction of it arriving on the earth eight minutes after fusion reactions on the sun generated the light.

In a peaceful Tuscan farm field, durum wheat plants basking in the sun in early June captured this energy from the sunlight through photosynthesis and used it to combine carbon dioxide extracted from the air with water extracted from the soil to produce the sugar glucose. The plants then released oxygen, a by-product of the reaction, into the air. Shortly thereafter, each wheat plant converted some of the glucose into starch and deposited it in the developing seeds. By late July, the wheat plants had matured into masses of plump wheat kernels suspended on dry ochre straw.

Italian farmers—many of them descended from the mitochondrial haplogroup H-bearing people who migrated from the north into Italy thousands of years earlier—repeated their ancestral traditions as they harvested the wheat seeds. A nearby mill ground the seeds into flour, and a processing plant turned the flour into lasagna noodles. An Italian chef in the restaurant expertly baked the noodles with tomatoes, basil, oregano, garlic, and cheeses in a terra-cotta oven to make a perfect lasagna.

As each bite of the lasagna passes over your taste buds, enzymes released in your mouth begin converting the wheat starch back into glucose. The process continues as your digestive system completes the breakdown of starch and transfers the glucose into your bloodstream. As you inhale with each breath, you extract oxygen from the air—a fraction of it perchance from the same nearby wheat fields that produced the wheat in the lasagna—and your lungs deposit the oxygen into your bloodstream. Your red blood cells contain hemoglobin—produced by genes in the globin clusters duplicated several times during our long evolutionary history—which carries the oxygen in the bloodstream. Cells throughout your body extract the glucose and oxygen from your blood, and, using aerobic respiration, they convert them back into carbon dioxide and water, releasing the sun's energy from the glucose for you to use. You expel carbon dioxide back into the air when you exhale. The water also returns to the environment, some of it carried on your breath, the rest of it eventually excreted from your body or evaporated from the surface of your skin.

In this example, we've focused on the interaction of two species—humans and wheat plants—with their environments. But the overall process is much more complex. The wheat plants required nitrogen to grow, which they extracted from the soil. Some of this nitrogen came from cattle manure spread on the fields by the farmers, perhaps from cattle who produced milk, which became cheese in someone's lasagna. The cattle obtained the nitrogen from clover they consumed in a pasture. The clover plants obtained their nitrogen from bacteria growing inside nodules on the plant's roots beneath the soil. These bacteria obtained the nitrogen from the air circulating around the soil particles.

The cattle relied on yet another type of bacteria in their digestive system to break down the plant material they ate to derive their own energy as they chewed their cud. These bacteria have a symbiotic relationship with the cattle. The cattle provide them with energy in the food they eat, and the bacteria, in turn, break down some of the fiber in the plants to produce sugars as an energy source for the cattle.

The plants consumed by the cattle, as well as the wheat plants used for lasagna, interacted with microscopic fungi in the soil to extract minerals such as phosphorus for their DNA and magnesium for their chlorophyll, the stuff that makes plants green. Bees and butterflies visited the clover flowers to obtain nectar, transferring pollen so that the plants could reproduce by making seeds. Insect pests, such as aphids, extracted nourishment from the wheat and clover plants while the aphids' predators, such as ladybird beetles, feasted on them.

The farmers, too, were an integral part of this complex interaction of the living with the nonliving. As they tilled the soil with their tractor-drawn plows, they altered the lives of countless microorganisms and invertebrates living within the soil, allowing some to thrive on the additional air in the loosened soil and causing others to die as their compacted-soil environment was disrupted. Weeds, which have naturally evolved over thousands of years to exploit tilled soil, grew in the farm fields and on the areas bordering the fields that had been disrupted by farming activity. The farm machinery ran on diesel fuel, refined from crude oil extracted from wells mostly in the Middle East, Africa, and the North Sea. The energy in this fuel was also a product of photosynthesis—a very ancient photosynthesis that happened in algae and plants that lived more than three hundred million years ago. This energy was trapped in their remains, which over eons of time were chemically transformed into crude oil. The farmers released the carbon trapped in those remains into the atmosphere as they burned the refined diesel fuel. We do the same every time we ride in a vehicle or heat our homes.

This example merely begins to scratch the surface of the complex interconnections of all life with the nonliving components of the world. As members of the earth's grand biosphere, we are increasingly altering the environment where we and all forms of life must thrive. Our species has been on this planet for about two hundred thousand years, and everywhere we've roamed, we have impacted the environment and the evolution of species around us. But massive *global* impact of humans is a recent phenomenon, confined to about the last thousand years or so, and especially to the past century. We are now impacting the evolution of life on the earth as never before. The reasons are complex, but one overwhelms and drives all the others—*the worldwide human population is exploding.*

MALTHUS REVISITED

In the previous chapter, we saw how the writings of Thomas Malthus captivated Charles Darwin. In them, he found a fundamental, underlying reason why natural selection could so powerfully drive evolution. To briefly recap, Malthus recognized that human populations have the potential to grow exponentially, but that in his day (late eighteenth century) food production was increasing at a linear, rather than exponential, rate. If these trends continued, Malthus surmised, the worldwide human population would grow too large for the food supply, and mass starvation would be the result. As noted in the previous chapter, Malthus did not foresee the effect scientific technology would have on food production, which grew exponentially along with exponential population growth.

Darwin saw in this idea how natural selection could work. He recognized that what Malthus foresaw for humans was true for *all* species. All have the potential for exponential population growth, and this potential is absolutely essential for natural selection to work. Although in nature we observe fluctuations in population sizes and occasional bursts or crashes, populations of most species remain relatively stable in size over long periods of time. The reason has to do with the complex interactions of organisms with one another and their environments, which prevent some, often most, individuals from reaching reproductive age and perpetuating their genes. As Darwin saw it, those who do survive and reproduce are likely to be those whose inherited characteristics offer them an advantage. They then transmit those characteristics to their offspring, and the species perpetually adapts to its environment whether the environment changes or remains stable.

For some species, the potential for exponential population growth is tremendous. For example, each year mature salmon from the oceans swim up streams to spawn, typically the same streams where they began their lives years earlier. Each female carries on average about two thousand eggs, which a male salmon bathes in sperm in the upper reaches of the stream. The female and male parents, exhausted from their mating run, die after reproducing, their fertilized eggs soon hatching into tiny fish no longer than your little fingernail. Their outlook for life is bleak. Most become food for other animals, who consume the vast majority before they can return to the ocean. Others succumb to disease or obstacles that block their migration. Those few who make it back to the ocean are still far from safety. A host of ocean predators, including people, consider adult salmon to be a great feast. Because there are so many threats to a salmon's life, those individuals who do reproduce must leave thousands of offspring to maintain the species. If the threats to their lives were removed, population sizes would explode within a single generation.

Although humans are not as reproductively prolific as salmon, our bodies devote a significant proportion of their biological activities and energies to reproduction. Human females begin producing mature egg cells between about nine and twelve years of age. Thenceforth, a female ovulates once (and rarely twice) per menstrual cycle until menopause, with interruptions during pregnancies and lactation. Human males constantly produce literally millions of sperm cells from about fourteen years of age often until death. Of course, only a tiny fraction of egg and sperm cells actually undergo fertilization and development to become a person.

Nonetheless, like all species, humans have the clear potential for exponential population growth. As a simplified hypothetical example, imagine that every generation of people leaves an average of four reproducing offspring for every female parent, a scenario not far from reality in some places on our planet. If this situation remains steady, populations should double in size with every generation—about twenty-five years for humans. A population of one million becomes two million after twenty-five years, four million after fifty years, eight million after seventy-five years, and sixteen million after one hundred years. At this rate, in one thousand years, the original population of one million will be about one quintillion people (a 1 with eighteen 0s following it), which is about 140 million times the size of our current world population of seven billion.

Although birthrates throughout most of human history have been high enough for human populations to *potentially* grow at a rate even greater than this, they never have grown at such a rate, at least on a worldwide scale. The reason is simple. People historically died from disease, starvation, childbirth, accidents, warfare, and a number of other causes at a rate sufficient to keep population growth in check. For most of human history, the world's popula-

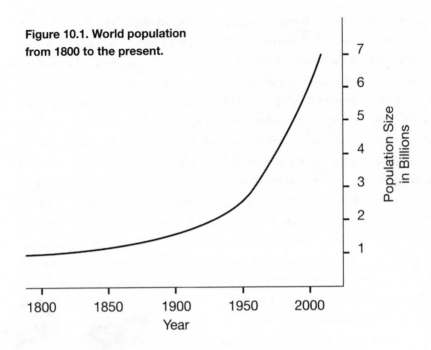

Figure 10.1. World population from 1800 to the present.

tion has grown ever so slowly. Only at the beginning of the nineteenth century did it reach one billion people, but from then on, the growth has exploded. A little more than a century elapsed until population doubled to two billion in 1927. Forty-seven years later, it doubled to four billion in 1974.[2] By 2011, thirty-seven years later, it had surpassed seven billion. The exponential growth trend is evident in figure 10.1.

What made this growth possible? Among the most important factors are scientific advances in medicine and agriculture. As we saw in chapter 8, death from infectious diseases and other medical issues has plummeted because of modern medicine and sanitation. Further, as we saw in chapter 9, scientific efforts to increase food production have been tremendously successful, allowing food production to increase exponentially at a rate greater than population growth. People who would have died from disease or starvation in times past are now alive—and reproducing.

Our current population explosion and our interaction with our environment will permanently influence the future of human evolution. With vastly greater numbers of people on the earth, the genes we all carry are more numerous and more varied than at any other time in human history. And our increased ability to migrate means those genes are spread more widely and recombine more readily with one another. Never before has our worldwide population been so diverse or so dispersed. And those diverse genes present in the billions of people alive today are the foundation for the future generations of humanity.

As archaeological evidence shows, we have impacted our environments for all of human history. But for most that time, the impacts have been localized. Enormous areas around the world, even entire continents, remained mostly free of human impact until recent times. Significant impacts began when humans became agriculturalists. They genetically altered wild plants and animals, domesticating them through artificial selection and, in the process, introduced an entirely new type of evolution—the emergence of new species through intentional and directed human effort. Agriculture permitted increased food production, allowing civilizations to emerge and populations to grow.

As the population data we just examined show, the worldwide human population began to dramatically climb early in the nineteenth century. This ramped-up growth coincided with an explosion in technological development. In fact, there is good evidence that technology made our population explosion possible. Technical advances in mechanized agriculture allowed farmers to increase food production and to convert additional land into productive farmland. Development of more sophisticated means of transportation offered the opportunity for people to spread throughout the world. And progress in healthcare and medical technology reduced the death rate, allowing more people to live, and to live longer.

Our increased reliance on technology also means that most people in modern times have a greater impact on local and global environments than in the past. Like me, most of you who are reading this book live in a prosperous society. You probably dwell in a house or apartment powered by electricity, rely on fossil fuel–powered vehicles for transportation, and consume food you did not grow yourself—some, if not all of it, transported over long distances. Compared to subsistence farmers in Africa, Asia, or Latin America, our impacts on the global environment are considerably greater, yet we often do not directly see them or think about them.

HABITAT LOSS DRAMATICALLY ALTERS ENVIRONMENTS AND EVOLUTION.

Today we can observe our planet in ways previous generations could only dream of. While writing this book, I've often turned to the Internet to view satellite images of various places on the earth. Much of the world is open ocean with little visual evidence of human intervention, although changes below the water's surface in ocean ecosystems as a consequence of human activities are well documented. And diminishing ice caps at the poles are a major sign of climate change. On the land, the extent of human impact is visually evident. Vast areas are farms planted with crops. Others are rangeland where domestic animals graze. Open-pit mines where minerals or coal are extracted appear as prominent features. Roads with multiple side-branches ending in circles, like cul-de-sacs, mark the locations of oil and gas wells. Artificial lakes, their dams visible as straight barriers, are common sites along rivers and streams. Throughout the landscape are cities and towns, interconnected with webs of roads and highways, railroads, electric lines, and pipelines.

These alterations of our planet's topography are essential for our world economy, and some, such as agriculture, are necessary for the survival of our current world population. They also displace natural habitats—the local environments in which native organisms live—and alter the course of evolution for the species who have evolved there. When these alterations are widespread throughout much of our planet—as many of them now are—the effect on evolution is global.

Farmland where crops are grown, rangeland where domestic animals are grazed, and forests managed for tree harvesting are often collectively categorized as agriculture. These agricultural practices occupy more land and alter more natural habitats than any other human activity. Large, highly mechanized commercial farms with conventionally grown crops produce much of the world's food, which is harvested in mass and processed, then shipped long distances. Current efforts to move toward more sustainable types of agriculture are essential, and we should do all we can to promote them. However, even if all our food were produced on small-scale farms with organic production, and if most of it were locally grown and sold, the earth's human population of about seven billion people would still require vast amounts of land to feed itself. As things now stand, we are and will be dependent on global-scale agriculture for our food, clothing, and building materials.

The nineteenth and twentieth centuries were not only a time of rapid human population growth, they were also necessarily a time of agricultural expansion. Broad prairie and forest lands, as well as wetlands and deserts, throughout the world were transformed into farmland and pasture. Forests were cut for wood then replanted or cleared for other uses. Nowhere is this transformation more evident than the Great Plains of the United States and Canada. When the worldwide human population reached its first billion in 1804, much of the land in this part of the world was native prairie interspersed with natural forests and wetlands. Great herds of American bison, pronghorn antelope, deer, and other large mammals roamed the land, freely feeding on the waving tall grass covering the seemingly endless prairie. Birds nested in the forests or on the ground, hidden in the grass or in the wetlands around rivers, ponds, and lakes, many of them migrating with the seasons. Rabbits, prairie dogs, mice, and other rodents also fed on native plants, as did countless insects. Predators, such as wolves, coyotes, foxes, hawks, and snakes kept the populations of plant eaters in check.

People were also a part of the equation then. The ancestors of Native Americans had entered that region between ten and fifteen thousand years ago. They thrived mostly on the native species, especially bison, for survival. They later adopted subsistence agriculture as domesticated corn, beans, squash, and other crops expanded northward from what is now Mexico. These people impacted their local environments, altering and becoming part of the ecosystems where they lived. But their impact was limited.

Today paintings, drawings, a few old photographs, and remnant parcels of land offer us images of what the midwestern United States and Canada looked like two centuries ago when the agricultural transformation was just beginning. What once was a great expanse of prairie is now covered with some of the world's most productive farmland and pasture. Fields of corn, soybeans, wheat, barley, oats, sorghum, sugar beet, sunflower, vegetables, alfalfa, tobacco, and other crops cover the land, stretching for miles on end. Cattle, sheep, horses, and other animals graze on pastureland. Patches of forest interrupt the farmland and in some places cover large areas. But even the forests are in most cases not native. Many of the trees now standing were planted after the old-growth forests were harvested for wood decades ago.

Throughout most of Europe, the United States, Canada, and other highly developed countries, nearly all land that can be productively used as farmland has been transformed for agriculture. In these parts of the world, farmland is actually on the decline as housing projects, commercial developments, road-ways, and utilities displace it.

Farmland is expanding, however, in other parts of the world. Nowhere is this expansion more rapid or successful than in Brazil, which is predicted to increase its agricultural production more than 40 percent within the coming decade.[3] South of the Amazon rain forest lies a large expanse of savannah called the Cerrado, which, during the past fifty years has been rapidly converted into massive commercial soybean and sugarcane farms. The soybeans are processed into a variety of products—vegetable oil, food additives, industrial materials, and animal feed—then sold on world markets. The sugarcane provides sugar exported throughout the world plus enough ethanol to fuel a majority of Brazil's automobiles, making the country essentially independent of foreign oil and an exporter of fuel ethanol.

In the nineteenth century, grazing land for animal production began an exponential expansion along with human population growth. Eighteenth- and nineteenth-century European landscape paintings depict cattle or sheep pleasantly grazing on serene pastureland. Romantic tales of cattle ranching across the western landscape dominated much of American popular literature, cinema, and television in the mid-twentieth century. These visual and lit-erary narratives reflect a transformative reality. On land where crops cannot be grown, large animals are often raised, feeding on native plants or increasingly on nonnative pasture grasses and legumes intentionally planted to make the land more amenable to large animal production. Enormous regions of range-land throughout the world have been transformed for animal grazing.

Also at the beginning of the nineteenth century, much of the forestland throughout the world had reached what are known scientifically as ecological climax communities, with tall old-growth trees, many of them more than a century old. Where human populations expanded, people cut the forests, harvesting the old-growth trees for valuable wood. If you hike through one of the many forests in temperate regions of the world, you are likely to see large rotting stumps of old-growth trees harvested years ago.

Worldwide, agriculture has displaced or altered native ecosystems, with immense effects on the evolution of native species. One of the most obvious is endangerment and extinction. Habitat displacement, along with unregulated and wanton hunting, has dramatically altered the evolution of animal species once prevalent. As one of the most famous and dramatic examples, the passenger pigeon was once a widespread and abundant bird in North America. Nineteenth-century accounts describe millions of birds flocking like clouds during seasonal migrations. Clearing trees from forests where the birds nested to make way for farm and pastureland, coupled with excessive hunting, drove the species to extinction by the early twentieth century. Accounts from the time tell of people saturating grain with alcohol to make the birds drunk and easy to capture and netting large numbers as people raided forests where the birds nested, killing them by the hundreds of thousands as they tended their eggs. The last individual of what had been billions of passenger pigeons died in 1914 at the Cincinnati Zoo.

The American bison nearly suffered the same fate. Settlers moving westward across the Great Plains of nineteenth-century North America wrote of enormous bison herds surrounding their wagon trains and extending as far as the eye could see, easy targets for people with rifles, who shot them for food or merely for sport. Excessive hunting, along with habitat destruction as grazing grounds became farmland or pasture, brought the American bison to near extinction. Only through the dedicated efforts of conservationists who established captive breeding programs was the species preserved. With its genetic diversity reduced to a mere fraction of what it once was, and with its habitat mostly displaced, the American bison's evolutionary path has taken a dramatic detour.

INDUSTRIAL EXPANSION IS IMPACTING EVOLUTION.

Like agricultural expansion, the Industrial Revolution coincided with the human population explosion. Fossil fuels, including coal, petroleum, and natural gas, became sources of abundant energy well beyond the capacity of renewable sources employed through much of human history, such as wood burning, water, wind, and draft animals. Coal provided energy and heat for factories, mills, electric plants, homes, and buildings. Crude oil refined into kerosene, diesel, and gasoline eventually became the fuels of choice for most vehicles, including automobiles, trucks, buses, trains, and airplanes. Where pipelines could deliver it, natural gas became an inexpensive, efficient, and relatively clean fuel.

Pollution has been a part of human environmental impact since ancient times. As evidenced in archaeological sites, local pollution from ancient garbage dumps, sewage, and wood burning predate written histories. Like ancient agriculture, the effect of these activities was restricted to local ecosystems during all the time when the global human population was relatively small. As most everyone in today's world is aware, pollution is an ongoing issue both locally and globally.

One of the earliest and best-documented examples of evolution by natural selection was directly related to air pollution, and it is an excellent illustration of how human activity can affect complex interactions of species in an ecosystem. This example includes a multifaceted interaction of bacteria, algae, fungi, trees, insects, birds, and people, but it is one that is easily understandable. We'll begin the story with the bacteria, algae, and fungi, which grow in association as lichens, which appear much like scaly, leafy, or fuzzy growths on tree bark and rocks. Lichens are highly sensitive to air pollution—so sensitive that scientists who study them have shown that they are effective bioindicators of air quality: lichens are usually among the first victims of consistently poor air quality.[4]

The Industrial Revolution in Britain relied heavily on coal combustion, which blackened buildings in Britain's cities. Nearby forests were also affected, and the polluted air began killing the lichens growing on tree bark. Speckled-white lichens that commonly grew on tree trunks and branches gradually disappeared in polluted forests. Trees that once appeared light turned dark as the lichens disappeared, exposing the tree bark to accumulating coal soot, which further darkened the trees.

Peppered moths are nocturnal insects, spending most of their active time flying at night. They rest during the day on tree trunks and branches, where birds feast on them whenever they find them. Before the Industrial Revolution, natural selection had favored moths whose wings had a speckled-white pattern resembling the lichens growing on tree bark. This type of speckled-white moth is called *typica* (figure 10.2). Typica moths were well camouflaged from birds when resting on lichen-covered tree trunks and branches and were more likely to survive daytime bird predation than moths of other colors. A rare genetic variant caused a few moths to be dark colored, a type called *carbonaria* (figure 10.2). In unpolluted areas, carbonaria moths were less likely to survive on a lichen-colored tree trunk because insect-eating birds could readily see them.

As the lichens died and trees became dark in polluted forests, the typica moths were increasingly exposed during the day, and birds began preying on them. The carbonaria moths, on the other hand, were now better camouflaged and better able to survive than the light-colored moths. Their numbers grew as the numbers of light-colored moths dwindled, a response to natural selection as the environment changed. Scientists readily noticed that carbonaria moths were the most predominant in polluted forests, whereas typica moths remained the most prevalent type in unpolluted forests.

What made this example so relevant was its extensive scientific documentation and a long series of criticisms and follow-up experiments. Noting the prevalence of different types in polluted and unpolluted forests, British

typica   carbonaria

Figure 10.2. Two types of peppered moth.

entomologist J. W. Tutt proposed in 1896 that bird predation was responsible for the high proportion of dark-colored moths. By 1924, the great British geneticist and mathematician J. B. S. Haldane had proposed mathematical genetic models based on natural selection to explain the change in prevalence from light- to dark-colored moths in polluted areas. In the 1950s, Bernard Kettlewell conducted a series of experiments in nature by capturing both typica and carbonaria moths, marking them, releasing them, and then measuring the rate of recapture in both polluted and unpolluted woodlands to document changes in the numbers of different-colored moths. As expected, he recaptured more carbonaria moths in polluted areas and more typica moths in unpolluted areas. He also photographed and filmed bird predation of moths during the day.

His experiments became the most famous textbook example of change in nature due to natural selection, but years later, his work fell under criticism, largely from opponents of evolution but also by some scientists. The scientists mostly questioned not whether the changes represented evolution but whether Kettlewell's experimental methods were appropriate and whether bird predation was really the cause. Some opponents of evolution, in an effort to discredit one of the so-called icons of evolution, were quick to claim that the change in these moths was not an example of natural selection, and a few even went so far as to accuse Kettlewell and other scientists of deliberate fraud.[5] As Michael Majerus, currently the foremost expert on peppered moth evolution, put it, "The exchanges have been acrimonious and have involved accusations of professional bullying, scientific fraud, data tampering, misquotation, misinterpretation born out of scientific ignorance, selective citation, and just plain lying."[6]

As it turned out, the claims of poor science had no basis, particularly those alleging fraud. Exhaustive recent experiments, led largely by Majerus, have confirmed in detail that natural selection driven by bird predation *is* responsible for changes in the moths and that changes in the peppered moth are well justified as a textbook example of how evolution in a natural ecosystem responds to human impact.[7]

Recent advances in industrial technology have done much to mitigate the effects of pollution. New and upgraded coal-powered electrical plants are capable of removing much of the noxious and particulate pollutants from the exhaust before it is expelled into the atmosphere. What remains is mostly carbon dioxide and water. Modern gasoline-powered vehicles have catalytic converters that convert certain pollutants into less harmful compounds before they exit the vehicle's tailpipe. Natural gas is a clean-burning fossil fuel widely used for home heating, electricity generation, industrial purposes, and even as an automobile fuel. Cars operating on natural gas are among the cleanest vehicles on the planet, producing exceptionally small amounts of noxious pollutants and less carbon dioxide than conventional vehicles, with fuel that costs less than gasoline. My family members and I have been driving natural gas vehicles since 1994 and have found them to be clean, reliable, and less-expensive alternatives to conventional gasoline-powered vehicles.

In spite of this progress, as world populations grow and world economies prosper, more people are turning to technologically advanced, more affluent lifestyles. With affluence and greater consumption of resources, each person's impact on the environment increases. The cumulative effect, even when advances in pollution control are taken into account, is an increase in overall pollution.

Of greatest concern in the long term is global climate change. Although politically disputed, there is little scientific dispute that climate change is under way and that increases in atmospheric carbon dioxide, methane, and other so-called greenhouse gases as a consequence of human activity contribute significantly to climate change.

Because ecosystems are so diverse and complex, the evolutionary consequences of climate change are only partially predictable. In the fossil record, we see evidence of what has happened in the past when climates changed. In the most severe cases, mass extinction of species blanketed the planet, followed by a resurgence of new species and ecosystems very different from those that preceded the change. Among the best-studied cases is the most recent mass extinction sixty-five million years ago, when a large meteor struck the earth and jettisoned massive amounts of debris into the atmosphere, causing major climate change. Among the many evolutionary changes precipitated by this cataclysmic event was the extinction of large dinosaurs and many non-flowering plants, opening the way for the emergence of mammals, insects, and flowering plants as predominant groups of organisms. As mammals, we owe our own evolutionary history to this event.

Although a natural event as catastrophic as this is still possible, it is highly unlikely. More frequent in evolutionary time are fluctuating climates, such as ice ages that come and go over periods of thousands of years. These events also impact evolution, resulting in species extinctions and the emergence of new species. It is quite telling that our nearest recent hominin relatives, the Neanderthals and Denisovans, went extinct within the past fifty thousand years, although not without leaving a smattering of their genes behind in some of us. The causes of their extinctions are not fully known. Climate change, disease, and competition with modern humans are all possibilities. Our closest living evolutionary relatives are chimpanzees, bonobos, gorillas, and two orangutan species that are now endangered by humans through hunting and encroachment on habitats. They, too, could suffer extinction in the not-too-distant future.

Endangerment of prominent animal species, such as the American bison, chimpanzees, gorillas, and orangutans attract public attention, causing us concern about how we manage our shared planet's resources. Less obvious are the small animals, invertebrates, plants, fungi, and microorganisms whose evolution is forever altered, and often halted, by human activity. It is best to think of our impacts on living organisms in terms of interconnected ecosystems and their evolution, instead of in terms of individual species. When one obvious species is threatened, we can be certain that other less obvious species are likewise subjected to evolutionary change and possible extinction.

SOME SPECIES DERIVE AN EVOLUTIONARY BENEFIT FROM HUMAN ACTIVITY.

Not all species are placed in peril by human actions. Life is so diverse and evolution is so resilient that some species derive advantages from human-caused changes in ecosystems and from the environments we create for ourselves. Several years ago, I was researching the evolution of sex determination—how nature makes females and males—in plants native to the American West.

My study site was a remote desert near Grand Junction, Colorado, and the plants were perennial shrubs with life spans of about ten years. Based on previous DNA studies we had conducted with these plants and observations by other scientists, I suspected that some of the plants were switching sexes from one flowering season to the next—a female one year might end up as a male the next, or vice versa. On a hot May day, when the plants were fully flowering, I examined hundreds of them, identifying them as females, males, or hermaphrodites (plants with both female and male flowers on the same plant). I affixed a harmless tag to each plant with its sex identified by an indelible mark. I planned to return the following year and record which (if any) had switched sex.

Dutifully, I arrived at the study site almost one year to the day from the time I had tagged the plants—and my heart sank. A herd of sheep had passed through the site and had eaten all the plants to the bare ground. My tags were scattered around, not a single one still on a plant, every spot of bare soil covered with sheep hoofprints.

The plants were perennials, and their root systems were still alive below the soil surface, so I hoped to find them regrown the next year, when I intended to salvage the study. For a third year I returned in May, and again my heart sank. The site was now covered with cheatgrass, a species native to the Middle East, accidentally introduced to the United States by humans with wheat seed brought from Europe. Now, years later, the site remains covered in cheatgrass. The native plants were unable to outcompete the nonnative cheatgrass and are now dead and gone. All that remains of them is their DNA stored in a laboratory freezer.

Cheatgrass is wildly successful as an introduced species in desert regions where human activity disrupts the land. This species exploits sites where the native plants have been removed by grazing, cutting, plowing, or fire, filling in the denuded land in a single growing season as it had on my study site. Evolution has treated this species well; it has a major advantage conferred by natural selection—cheatgrass thrives on wildfires. Seeds germinate in the spring, watered by spring rains or melting snow. By late May and early June, the grass is lush and green, topped by long, wispy seeds. In late June, the grass turns dry like straw, dropping its seeds to the ground. Millions of seeds remain in the soil for years, dormant, with just a fraction germinating each spring, yet still enough to fill a barren spot of land with cheatgrass. The dry cheatgrass is a serious fire hazard from midsummer to late fall. When it ignites from a lightning strike, a discarded cigarette butt, an unextinguished campfire, a spark, or even the hot catalytic converter of a vehicle passing over the tinder-dry grass, it quickly burns with an intense heat that kills most native plants. The heat also stimulates the cheatgrass seeds lying dormant in the soil to germinate the following spring, where they exploit the site stripped of the native plants that died in the fire.

Enormous tracts of land in the American West are now covered with cheatgrass, many of them in places burned in wildfires. A single fire fueled by cheatgrass may cost tens to hundreds of millions of dollars in suppression costs, lost resources, and rehabilitation.[8] Cheatgrass is now evolving in its new home in North America. Unless we find a way to eradicate it, which seems unlikely, cheatgrass in America, now separated from its ancestral types in the Middle East, will eventually evolve into a new species adapted to the American West.

The majority of introduced species do not survive well in their new environments. They are usually best adapted to their native habitats, their genetic structure having been shaped through natural selection for the place where they originally evolved. A significant fraction of introduced species, however, thrive in a new environment where they are already adapted to the climate and have the added advantage of being free of natural diseases, predators, or other factors that limit their population growth. Some, like cheatgrass, may become invasive, spreading widely as they displace native species.

Humans have introduced most invasive species, some intentionally, others unintentionally. One of many examples is the zebra mussel. This clam-like freshwater species is native to the Caspian and Black Seas in Central Asia. The adults are small, clam-like animals that attach themselves to solid substrates, including boats. Adults can survive outside of water for up to weeks at a time, often attached to a boat transported from one body of water to another. Throughout the latter part of the twentieth century, zebra mussels spread into freshwater rivers and lakes in Europe as people carried boats from one body of water to another. They were found in the North American Great Lakes in 1988, probably introduced in ballast expelled by ships traveling the St. Lawrence Seaway, and are rapidly spreading on pleasure boats when owners fail to thoroughly clean and dry their boats. The mussels displace native species and are costly as they clog intake pipes for water supplies, gates, pumps, and hydroelectric plants. A major campaign is under way in the United States to put the brakes on their spread.

Hundreds of other invasive species threaten the survival of native species and the economic activities of people worldwide. The United States Department of Agriculture provides up-to-date information on the most threatening invasive species and strategies for their control on their National Invasive Species Information Center page at http://www.invasivespeciesinfo.gov.

Some species derive an evolutionary benefit from their close association with us. One of the oldest and most common of these beneficial relationships is the one we share with the house mouse. The mouse evolved in Europe but quickly spread throughout the world during the European colonial expansion as mice hitched rides on ships, in food containers, and in luggage. Although house mice are wild animals evolved from field mice, they have become adapted through natural selection to live in close association with humans in environments we unwittingly create for them. In several ways, we have influenced their evolution. Our habits of living in buildings and storing food provided an environment in which they could evolve. To control them, scientists introduced a rodenticide called warfarin, which is still widely used in poisoned bait to kill mice—and at low doses as a drug to prevent blood clotting in humans—although warfarin-resistant strains of mice have been evolving since the 1960s.[9] More than a century ago, scientists captured wild mice and altered their evolution by breeding domesticated strains, especially white albino and hairless types, which are used as laboratory animals for research and are sold as pets.

Human parasites also have evolved to thrive in environments where people live in close quarters. For example, the common bedbug is a blood-sucking insect that first evolved in the Middle East and later spread throughout most of the world as humans inadvertently carried them in their belongings while traveling. After decades of effective control with pesticides during the early twentieth century, bedbugs are reappearing with a vengeance, especially in apartment buildings and hotels where they can be readily transported from one place to another in contaminated laundry, furniture, or suitcases. Large populations of people, increased travel, and recently evolved resistance to pesticides through natural selection are factors favoring their resurgence.[10]

The same is true for other pests. As mentioned in chapter 7, mosquitoes that carry the malarial parasite rapidly became resistant to DDT. Officials administering mosquito control efforts turned to other pesticides, and once again natural selection took over. Resistance to other pesticides is now evolving in parasite-bearing mosquitos.[11]

WHY DOES IT MATTER?

As beneficial wild species have been lost or threatened, such as the passenger pigeon and American bison; as introduced invasive species have spread to most places on the earth where people live; and as pests continue evolving to thrive in the environments we create for ourselves, managing these situations is a high public priority. As a consequence, government agencies regulate land use to protect natural ecosystems and wildlife in designated areas. Environmental impact studies are required when developments may impact local ecosystems, and they influence decisions regarding land development. Laws, such as the Endangered Species Act in the United States, protect species identified as being in peril of extinction. Fishing and hunting, including commercial ocean fishing and commercial harvest of wildlife, are regulated to manage and protect wild species we use for food. Certain areas of land are set aside as natural areas or wilderness and managed as natural ecosystems. Abandoned mines and farmland are often reclaimed through restoration of native ecosystems. Scientists and environmental advocates monitor ecological and evolutionary changes to provide up-to-date and accurate information to government agencies and those responsible for management and legislation. To be successful, all these activities require oversight by experts who have a solid understanding of evolution and how it operates in nature. Let's look at a couple of examples to see why.

The first example is the story of paclitaxel, a potent drug used to treat several types of cancer. The Pacific yew is a slow-growing, small tree native to the Pacific Northwest region of the United States and southwestern Canada, where it grows naturally among other trees in forests. For many years, loggers considered it to be a weed tree of little value for wood. When logging companies harvested forests in this region, then replanted them, they did not replant Pacific yews, opting instead for economically valuable lumber species, such as Douglas fir. As a consequence, Pacific yew populations declined.

In the early 1960s, under a National Institute of Cancer program, scientists found that extracts from the bark of Pacific yew trees showed potential for treating cancer. The chemical compound responsible for the potential anticancer effect was purified and named taxol. It turned out to be so chemically complex that artificial synthesis was not economically possible. Its only known source at the time was the bark of the Pacific yew, and it was present in excruciatingly small quantities—only about 0.0004 percent in the bark.

By the late 1970s, research showed that taxol was indeed an effective anticancer drug in laboratory mice, and clinical trials in humans began in 1984. By 1988, the results from clinical trials made it clear that taxol was a powerful anticancer drug in humans and that demand for it would be high once it was approved and commercially released. But there was an enormous problem—there were not enough Pacific yew trees in nature to treat even a fraction of the cancer patients who needed taxol, and the tree grew so slowly that it could not be readily cultivated for commercial production. Environmentalists and forest managers raised immediate concerns about conserving existing trees; many had already been sacrificed for their bark during the clinical trials.

Through controversial negotiations, Bristol-Myers Squibb received an exclusive contract from the US federal government to develop and market taxol (now trademarked as TAXOL®), and gave it the generic name paclitaxel. Initial commercial marketing began with paclitaxel isolated from Pacific yew bark but with a commitment to develop a new source. Scientists began searching for alternative sources of paclitaxel, looking first in the evolutionary relatives of the Pacific yew. They found a candidate in the European yew tree, which contains a compound similar to taxol that could be harvested from the tree's needles without seriously harming the trees. Scientists developed a chemical method for conversion of this compound from European trees into paclitaxel, and commercial production began in Europe. By 1995, reliance on Pacific yew bark had ceased, and the tree was no longer in serious danger. In 2004, scientists implemented a new technology that relies on plant cells cultured in large vats where the cells produce paclitaxel directly, requiring no material whatsoever from yew trees.[12] For this discovery and its commercialization, Bristol-Myers Squibb received the 2004 Greener Synthetic Pathways Award from the US Department of Energy.[13] Paclitaxel is now a successful generic drug used for cancer treatment.

A second example illustrates how observation of evolution in nature can lead to beneficial innovations. For a time, I shared my research laboratory with Howard C. Stutz, a brilliant evolutionary geneticist who pioneered several applications of evolution for the benefit of society. One of his innovations was a successful evolutionary approach for restoring lands disrupted by strip mining. Coal consists of ancient and chemically transformed remains of algae and plants that lived during the Carboniferous Period between 359 and 299 million years ago. The material was deposited as sediment, so coal is now often discovered in vast horizontal layers beneath the land surface. When the land is relatively flat, and the coal layers are not too far beneath the surface, strip mining is often the most efficient method for extracting the coal. To initiate a strip mine, a mining company excavates a strip of land a few hundred feet wide and as long as mile or two. Enormous machinery removes the soil, gravel, and rock, called overburden, until the coal layer is exposed. The company then excavates the coal from this enormous pit and ships it off to be combusted.

After the coal has been removed from the first strip, the mining company continues excavating a new strip adjacent to the first one, backfilling the first strip with the overburden from the new strip. When the coal has been removed from the second strip, the company excavates a third strip, then a fourth, and so on, each time backfilling the previous strip with overburden from the new one. Eventually, the process may upend thousands of acres of land, displacing all plants and animals that were originally there, profoundly disrupting the local environment.

In the United States, strip mining for coal is heavily regulated. Mining companies must restore the land when the mining operation is complete. The companies save the topsoil and replace it on the surface after a completed strip has been backfilled. They then revegetate the area, often with native plants. Here is where an understanding of evolution is essential. Although the mining sites have been covered with topsoil, the underlying soil structure and topography are not the same as they were before mining. Plants that once grew there may not successfully grow on the disrupted site, and failed revegetation can be an environmental disaster.

Throughout much of his career, Stutz had studied natural hybrid swarms—populations of plants that arise when two genetically distinct types of plants come into contact and mate with each other. Their offspring, which form the hybrid swarm, are highly diverse, much more diverse than most native populations. In one case he studied, two related species had met and produced a highly diverse hybrid swarm, and later an interstate highway was built across the swarm. He discovered a new species that had evolved from this hybrid swarm, which then colonized the construction-disturbed land in the highway median for several miles.[14]

This observation in nature gave him an idea. He surmised that if scientists hybridized genetically diverse types of plants to artificially create a hybrid swarm on land that had been strip mined, the probability of plants emerging from the genetically diverse hybrid swarm and successfully colonizing the disturbed land would be increased. The idea was straightforward and purely Darwinian. In *The Origin of Species*, Darwin repeatedly emphasized that natural selection was dependent on inherited variations, and the greater the variation, the greater the chance that successful types would evolve. Stutz tested his idea on strip-mined land and found it to be a resounding success. Land once denuded by mining could be successfully restored with genetically diverse native plants in an artificially produced hybrid swarm, as natural selection favored those plants best adapted to the disturbed sites.

There are hundreds of similar examples in which scientists have applied evolutionary principles and technology to protect environments and benefit humanity. The paclitaxel example also shows how seemingly unimportant species, some of which are in evolutionary decline because of human activity, can benefit humanity. Habitat decline endangers and may even drive to extinction species that share the earth with us, and some of those species may carry yet unknown benefits for us.

What can we do to ensure the future of living organisms that have evolved with us on our shared planet, and in the process ensure our own future? The world's human population continues to grow, but the growth rate is declining. Eventually the growth rate will stabilize and overall growth will cease, but not before several additional billion people have joined our global family. The demands on our planet's resources, especially for food and energy, will continue to rise. Technological development in agriculture, energy, and other areas will increase efficiency but, in all likelihood, not nearly enough to offset increased demands. Instead, individually and collectively, we will need to adopt more sustainable lifestyles, especially those of us who live in affluent countries, where we consume a disproportionate amount of the world's resources.

EPILOGUE

The evidence of ancient and ongoing human evolution is overwhelming. The examples in this book are merely a select few of thousands that fill volumes of scientific journals, and the number of examples continues to grow as evolutionary biologists publish new discoveries. Beyond the evidence of our own evolution, we've seen how an understanding of our place in evolution helps improve our health, feed our growing population, and protect our environment. These are all *practical* reasons for studying evolution. They illustrate why evolution matters for the future of our planet and for our quality of life.

There is another reason why evolution matters—*it helps us understand who we are*. It explains the intricacies of our unique anatomy and how our bodies function. Through it, we learn why we get sick, how our immune systems work to battle infectious disease, and how pathogens overcome our efforts to control them. It shows us how our DNA evolved into an enormous and complex genome, organized into chromosomes with genes, and littered almost everywhere with millions of invaders and freeloaders. The patterns of inheritance that govern DNA transmission through countless generations are inextricably tied to evolution and underlie evolutionary change in both the short and long terms.

And it's not just us. Evolution explains the whole of life, from the atoms, molecules, and cells that compose every organism that now lives or has ever lived to the myriad adaptations that allow diverse organisms to interact with one another and with their environments. The fossil record, the anatomy of living beings, the geographical distributions of organisms, and the vast stories written in DNA increasingly allow us to unravel the magnificent history of life.

Evolution is not just the central theme of biology but an integral theme for science in general. Astronomy, physics, chemistry, geology, and mathematics are all related to evolution. For instance, advances in astronomy have

279

given us a far-reaching vision of our universe. No longer do we view our planet as the center of the universe as people did a few centuries ago. Instead, we now understand much about how our solar system formed and how conditions on the earth made possible the evolution of life.

Physics is essential for understanding evolutionary history. Nuclear physics, which includes radioactive decay, is a key component for dating fossil remains. And other branches of physics are important for evolutionary studies, such as determining whether bone structures revealed in fossils are consistent with the forces the bones must bear for particular types of posture or locomotion—for instance, habitual upright bipedality in ancient hominins.

Chemistry helps us understand DNA, RNA, and protein and how they evolve. For example, the chemistry of DNA tells us that certain types of mutations, called *transitions*, are more likely to occur than other types of mutations, even though transitions constitute a minority of all *types* of mutations, albeit not a minority in the total *number* of mutations. When we examine mutations that have appeared throughout evolutionary history, we find more transitions than any other type of mutation, exactly as chemistry predicts. As another example from chemistry shows, flowers often smell sweet because natural selection has favored the production of volatile chemical scents to attract insect pollinators. And the bright colors of those flowers are chemical pigments that plants make to visually attract those insects. Evolution explains the complex pathways plants use to produce these chemicals. And chemists use what they learn from chemical evolution to synthetically produce similar chemicals not found in nature.

Geology is intricately tied to evolution and underlies much of what we know about distant evolutionary history. Continental drift coupled with uplift of mountain ranges explains why we find fossils of marine organisms on mountains, or fossils of species adapted to tropical climates in places now near the poles. Earth's long history, as revealed through geological science, tells us when and where certain types of life lived and died. Geologists use this evolutionary information to find the most likely places where deposits of oil, coal, and natural gas are located.

Math is the language of science, and it explains a multitude of patterns and statistical probabilities associated with evolution. Expert mathematicians, among them J. B. S. Haldane, Sir Ronald A. Fisher, and Sewall Wright, derived equations to describe how natural selection works, and experiments have confirmed many of their theoretical equations. Evolutionary scientists routinely apply math to validate their studies; nearly all scientific articles on evolution contain mathematical and statistical analyses.

Right now is the best time in history to be a scientist. Advances in all areas of science, biology in particular, have never been so rapid, so informative, or so relevant. Passion for scientific discovery is palpable in the laboratories of academia, government, and industry, and in classrooms from grade school to graduate school.

However, in the midst of this passion and rapid pace of discovery, a looming cloud hangs over science education, particularly when it comes to evolution. I belong to a group of professors who travel throughout our home state and to various places in the United States conducting workshops for high school science teachers on how to teach evolution.[1] No matter where we go, we find teachers who are emotionally torn between their bright optimism and intense frustration. Optimism because they understand how exciting modern evolutionary science is and how important it is for society. Frustration because they face stiff opposition from students, parents, community groups, legislators, government officials, and school administrators when they teach evolution.

Two factors fuel this opposition. First, about 40 percent of people in the United States reject human evolution outright, according to reliable scientific polls conducted by Gallup, the Pew Research Center, and other respected polling organizations.[2] As a small bright spot in the cloud, the Gallup poll on evolution has been repeated ten times since 1982, and the most recent poll (December 2010) showed that 40 percent of Americans reject evolution, the lowest proportion in the poll's history.[3] Although consistently a minority, this proportion remains surprisingly large, especially in light of the vast amount of evidence supporting human evolution.

Second, through a well-funded effort by powerful political organizations, opponents of evolution have attempted to eliminate or suppress evolution education, in spite of court decisions (including several by the US Supreme Court) that affirm the scientific nature of evolution and the nonscientific stance of its so-called alternatives. One of the most recent major legal decisions was the 2006 *Kitzmiller v. Dover Area School District* decision, which determined that "intelligent design" was not a scientific alternative to evolution and that teaching it in public schools violated the establishment clause of the First Amendment to the US Constitution. The decision, authored by Judge John E. Jones III, is thoughtfully written and well worth reading. It is freely available online at http://www.pamd.uscourts.gov/kitzmiller/kitzmiller_342.pdf.[4]

The most recent approach by groups opposing evolution is an appeal to so-called academic freedom, claiming that teachers should be allowed to "teach the controversy" and by so doing instill doubt in students' minds regarding the scientific validity of evolution.[5] Although evolution may be politically controversial, there is no controversy among scientists about its validity—the evidence supporting it is so substantial that its reality cannot be reasonably disputed. A very small minority of scientists choose to dispute it, and it is evident from their writings that most of them deny evolution for religious rather than for scientific reasons.[6]

Although science, current law, and the establishment clause of the US Constitution's First Amendment support them, many American science teachers find themselves under intense scrutiny when they teach evolution. To protect themselves, some avoid it altogether, whereas others dilute it. When they do, their students fail to learn the central theme of modern biology and one of the most essential components of modern science, contributing to an appalling lack of scientific literacy.[7] Many teachers, however, appropriately teach evolution, choosing to weather the inevitable criticism. By contrast, some parents send their children to private or charter schools, or homeschool their children, where they can be certain that evolution is either excluded or disdained.

The underlying reason for opposition to evolution, especially human evolution, is clear. The evidence that we evolved runs counter to young earth creationism—the idea that God created the universe, the earth, and life (including humans) within the past ten thousand years. The Gallup poll mentioned earlier is especially informative. As we saw in chapter 1, it asked respondents to choose one of three options:

1. Human beings have developed over millions of years from less advanced forms of life, but God guided this process.

2. Human beings have developed over millions of years from less advanced forms of life, but God had no part in this process.

3. God created human beings pretty much in their present form at one time within the last 10,000 years or so.[8]

Option 1 is often called *theistic* evolution because it recognizes the validity of human evolution but implies a role for God in the process. Throughout the poll's history, since 1982, the proportion of Americans who agree with this option has remained consistent, averaging 37.5 percent; in the most recent poll, from December 2010, it remained steady at 38 percent.

The other two options, however, have undergone a slight change in the most recent poll. Option 2, which can be classified as *naturalistic evolution*, has remained relatively low, between 9 and 11 percent until 1999, followed by a gradual increase to 16 percent in December 2010.

The largest proportion of Americans agree with option 3, classified as young earth creationism or special creationism, which clearly implies rejection of evolution. It remained steady between 44 and 47 percent until 2007, when the proportion began a slow decline to its most recent level of 40 percent. In spite of this slight decline, it remains the most popular of the three options. For details and graphic representations of this poll, see http://www.gallup.com/poll/21814/evolution-creationism-intelligent-design.aspx.

Several conclusions from this poll are readily apparent. First, a large majority of Americans are religious, believing that God guided our creation (78 percent in the most recent poll, combining options 1 and 3). Second, this majority is almost evenly divided between theistic evolution (option 1 at 38 percent) and young earth creationism (option 3 at 40 percent). Third, a slight majority accepts human evolution, be it theistic or nontheistic (54 percent, combining options 1 and 2), whereas a substantive minority reject it (40 percent agreeing with option 3). With such a strongly divided public, tension over evolution is inevitable.

This tension derives from the perception that teaching evolution is a threat to religious faith. Evolution is absolutely incompatible with young earth creationism, and as students who believe in young earth creationism learn about evolution, they must change their religious views or choose to reject the evidence of evolution. A significant proportion, many during their high school or college years, migrate toward theistic evolution, whereas others adhere to young earth creationism. A small minority abandon religious beliefs altogether.

Although scientists who write about evolution profess a wide range of religious beliefs, or lack thereof, they are essentially unified when it comes to evolution. Two polar examples are Richard Dawkins and Francis Collins. Dawkins is a well-credentialed evolutionary biologist; until his retirement in 2008 he held the Simonyi Professorship for the Public Understanding of Science at the University of Oxford. He is a prolific author of some of the world's most popular books; among his most recent are *The Greatest Show on Earth: The Evidence for Evolution*, and *The God Delusion*, both international best-sellers. He was raised as an Anglican but is now widely known as "the world's most prominent atheist." Collins, by contrast, characterized himself as an atheist during graduate school but converted to evangelical Christianity upon seeing the faith of people on their deathbeds during his medical training, and as a result of personal religious experiences. He was director of the Human Genome Project until his appointment as director of the National Institutes of Health in the United States, the post he currently holds. Pope Benedict XVI honored him in 2009 by an appointment to the Pontifical Academy of Sciences. His book *The Language of God* is also an international bestseller, promoting the compatibility of science and religion, including evolution.

Dawkins and Collins, both famous and highly respected biologists, could hardly be farther apart on the religious spectrum, but they are in complete agreement when it comes to the reality of evolution, including human evolution. The same can be said of the overwhelming majority of scientists, among them thousands of researchers, who augment day-to-day the body of scientific knowledge with their discoveries, and teachers who collectively share their passion for science with millions of students. I am proud to count myself among them as a teacher and researcher. Their views on religion diverge considerably, but their acceptance of the powerful evidence of evolution is nearly unanimous.[9]

What really matters is not scientific consensus itself but the reason for it, which is the powerful and abundant evidence supporting evolution. To counter evolution with religiously motivated "controversy" in public science education is deceptive to students and, as courts have repeatedly determined, is a violation of the US Constitution's establishment clause. The scientific evidence of evolution is not necessarily a serious threat to religious faith, as is claimed by the significant proportion of Americans who accept theistic evolution and the small minority who reject religion. Even Darwin felt compelled to comment near the conclusion of *The Origin of Species*, "I see no good reason why the views given in this volume should shock the religious feelings of anyone."[10] But even if evolution *were* a serious threat to religion, scientists would still have an obligation to draw conclusions from the evidence, and the evidence clearly and overwhelmingly tells a magnificent story of our own evolution coupled with that of all other organisms on the earth. And from this evidence, we are better able to improve conditions for humanity, and, more importantly, glean a fuller understanding of who we really are and our place in the great expanse of life.

Yes, we evolved and are evolving, along with the rest of life, and it matters—now more than ever.

ILLUSTRATION CREDITS

CHAPTER 2

Image of Albert Einstein redrawn from a 1905 photograph of Albert Einstein at the Bern Patent Office from the online archives of the Historisches Museum Bern, http://www.bhm.ch/downloads/04_Einstein_in_Bern.jpg. Original photo by Lucien Chavan from the Albert Einstein Archives in Jerusalem.

Figure 2.1. Composited from the twentieth US edition of *Gray's Anatomy* (1918).

Figure 2.2. From Ernst Haeckel, *Kunstformen der Natur* (Leipzig: Bibliographisches Institut, 1904).

Figure 2.3. Composited from the twentieth US edition of *Gray's Anatomy* (1918).

Figure 2.4. Composited from skeletal specimens.

Figure 2.5. Human limb redrawn from Paul Marie Louis Pierre Richer, *Nouvelle Anatomie Artistique du Corps Humain* (Paris: Pion-Nourrit et cie, 1906). Other drawings composited from skeletal specimens.

Figure 2.6. Adapted and redrawn from W. Ellenberger, H. Dittrich, and H. Baum, *Handbuch der Anatomie der Tiere für Künstler* (Leipzig: Theodore Weicher, 1901).

Figure 2.7. Drawn from museum specimens in the Smithsonian National Museum of Natural History, Washington, DC.

Figure 2.8. Whale skeleton and horse limb drawn from museum specimens. Dog's limb adapted and redrawn from W. Ellenberger, H. Dittrich, and H. Baum, *Handbuch der Anatomie der Tiere für Künstler* (Leipzig: Theodore Weicher, 1901).

Figure 2.9. Human embryo and auricula muscles redrawn from the twentieth US edition of *Gray's Anatomy* (1918). Goosebump, eye, and ear drawn from life.

Figure 2.10. Adapted and redrawn from Paul Marie Louis Pierre Richer, *Nouvelle Anatomie Artistique du Corps Humain* (Paris: Pion-Nourrit et cie, 1906), and composited from X-ray images of human tails.

Figures 2.11, 2.12, 2.13, and 2.14. Adapted and redrawn from Paul Marie Louis Pierre Richer, *Nouvelle Anatomie Artistique du Corps Humain* (Paris: Pion-Nourrit et cie, 1906), and from skeletal specimens.

Figure 2.15. Adapted and redrawn from Paul Marie Louis Pierre Richer, *Nouvelle Anatomie Artistique du Corps Humain* (Paris: Pion-Nourrit et cie, 1906), and from W. Ellenberger,

H. Dittrich, and H. Baum, *Handbuch der Anatomie der Tiere für Kunstler* (Leipzig: Theodore Weicher, 1901).

CHAPTER 3

Figure 3.1. Drawn from fossil specimens.

Figure 3.2. Drawn from a museum replica, Utah Valley University, College of Science and Health.

Figure 3.4. From C. Darwin, *Origin of Species*, 6th ed. (1878), and from Darwin's notebook, http://upload.wikimedia.org/wikipedia/commons/4/4b/Darwins_first_tree.jpg.

Figure 3.6. Adapted and redrawn from Paul Marie Louis Pierre Richer, *Nouvelle Anatomie Artistique du Corps Humain* (Paris: Pion-Nourrit et cie, 1906), skeletal specimens, and a museum replica, Utah Valley University, College of Science and Health.

Figures 3.7, 3.9, 3.10, and 3.13. Drawn from museum replicas and composites of published and sculptural reconstructions.

Figure 3.14. Composite of previous figures.

CHAPTER 4

Figure 4.1. Redrawn after a fifteenth-century Italian engraving in the National Gallery of Art, Washington, DC.

CHAPTER 5

Figure 5.1. Banding patterns redrawn from the National Center for Biotechnology Information (NCBI), http://www.ncbi.nlm.nih.gov.

Figures 5.2, 5.3, and 5.4. Adapted from D. J. Fairbanks, *Relics of Eden* (Amherst, NY: Prometheus Books, 2007).

CHAPTER 7

Figure 7.1. Adapted from resources of the National Center for Biotechnology Information (NCBI), http://www.ncbi.nlm.nih.gov.

CHAPTER 8

Figure 8.1. Drawn from a flask in the Pasteur Museum, Paris.

CHAPTER 9

Figure 9.1. Photograph by Daniel J. Fairbanks.

Figure 9.6. From http://commons.wikimedia.org/wiki/File:Vincent_Van_Gogh_-_The_Potato _Eaters.png.

CHAPTER 10

Figure 10.1. Derived from data in United Nations Population Division, *World Population Prospects: The 2000 Revision*, Vol. 3 (New York: United Nations, 2000), p. 155.

Figure 10.2. Drawn from museum specimens.

GLOSSARY

admixture: Mating of two individuals whose genetic backgrounds are substantially different.

ancestral feature: An anatomical structure or DNA sequence that is present in evolutionary ancestry, as opposed to a derived feature, which has changed from the original form or sequence.

antibiotic: Any one of a large number of chemical substances, often produced naturally in fungi, that kills bacterial cells or inhibits their growth.

artificial selection: The intentional action by humans, usually in breeding plants and animals, of determining which individuals mate and reproduce with the intent of improving inherited characteristics in the offspring.

atavism: Appearance of an ancestral anatomical feature, such as a true tail in humans, that is absent in most individuals of the species but was present at one time in the species' distant evolutionary ancestry.

australopithecine: A grouping of hominin species that lived from approximately four until two million years ago in Africa and probably included species ancestral to the genus *Homo*, from which modern humans evolved. Australopithecines were smaller in stature, had relatively small brains, and had habitual upright bipedalism when compared to modern humans.

bacterium (plural: bacteria): Single-celled microorganism that has a relatively small circular DNA molecule as its genetic material and does not have a cell nucleus. Many human pathogens are bacterial.

Beringia: A wide land bridge between Siberia and Alaska that existed during the most recent ice age where the Bering Strait currently is. Humans migrated across it about fifteen thousand years ago to populate the American continents.

biological species concept: A concept defining a species as a group of individuals who are able to mate with one another to produce fully fertile offspring.

bipedalism: The ability to walk on two feet.

breed: A group of individuals within a species that have a defined set of inherited characteristics, and those characteristics are maintained in the offspring when members of the same breed mate.

central concept of molecular biology: The concept that a gene in DNA encodes an RNA molecule through transcription, which, in turn, encodes a protein through translation.

chromosome: A single DNA molecule in cellular organisms bound to and stabilized by proteins.

chromosome fission: Breakage of a single chromosome into two functional chromosomes.

chromosome fusion: End-to-end fusion of two chromosomes to form a single functional chromosome with a single DNA molecule.

chromosome inversion: Rearrangement of a segment within a chromosome so that its orientation is inverted relative to the rest of the chromosome.

chromosome translocation: Transfer of a chromosomal segment from one chromosome to another so that the segment becomes part of the recipient chromosome.

clade: A set of organisms that share a common ancestor, grouped according to degree of relatedness.

comparative anatomy: Comparison of anatomical structures in related species to determine how those structures evolved from common ancestry.

consensus sequence: The most representative DNA sequence derived from alignment of similar sequences from different individuals of the same or related species.

cranium: The portion of the skull containing the upper jaw and braincase, not including the lower jaw.

creationism: A movement promoting the belief that God created all forms of life intentionally, separately, and individually, usually excluding evolution as the mechanism for creation.

crossover: An exchange of chromosome segments, usually between maternal and paternal copies of the same chromosome.

derived feature: An anatomical structure or DNA sequence that has recognizably diverged from its original structure or sequence, as opposed to an ancestral feature, which has remained essentially the same as its original form or sequence.

DNA: Deoxyribonucleic acid, a linear molecule composed of nucleotides, which contain the bases T, C, A, and G, is the inherited material of all cellular organisms.

domestication of plants and animals: The evolution of distinct species of plants and animals as a direct result of artificial selection by humans for the purpose of providing a benefit to humans.

dominant variant: A variation in DNA that is expressed in the outward appearance of an organism when present on at least one of the maternal and paternal copies of chromosomes.

early hominin: A grouping of hominins that lived from approximately seven until four million years ago, from the time of the human-chimpanzee divergence until the emergence of australopithecines.

evolution: Genetic change over many generations that ultimately results in the emergence of new and different species from a single ancestral species.

exon: A segment of DNA transcribed into RNA that ultimately remains in the final version of RNA encoded by a gene.

fission: *See* chromosome fission

foramen magnum: The opening in the cranium where the brain stem exits the cranium and connects to the spinal cord.

fossil: The anatomically identifiable remnant of an ancient organism, usually preserved in sedimentary rocks.

fusion: *See* chromosome fusion

fusion gene: A gene derived from the end-to-end fusion of two or more segments of different genes.

G-banding pattern: A barcode-like pattern of alternating light and dark bands on a chromosome that appears when the chromosome is stained with giemsa.

genetic drift: The random change from one generation to the next of the proportion of individuals carrying a particular genetic variant. Genetic drift ultimately results in either the complete loss or complete fixation (uniform presence in every individual) of a genetic variant.

genetic erosion: A loss of overall genetic diversity in a species, most commonly identified in domesticated plants and animals due to intensive breeding efforts.

genome: The complete collection of all DNA sequences in one chromosome for bacteria, or one full set of chromosomes for organisms with a cell nucleus. Most multicellular organisms (such as humans) have two genomes in each of their cells, one inherited from each parent.

genus: A grouping of closely related species that have recent common ancestry. For example, horses and donkeys are different species but belong to the same genus (*Equus*).

germline: The cells that contribute to the next organismal generation, including egg and sperm cells and their precursor cells.

giemsa: A chemical stain that produces alternating light and dark bands on chromosomes, determined by relative proportions of A-T and G-C base pairs in the DNA.

haplogroup: DNA molecules grouped by the set of common ancient variants they carry, usually applied to mitochondrial and Y-chromosome DNA.

haplotype: DNA molecules grouped by a common set of all DNA variants they carry, both ancient and modern.

HERV: *See* human endogenous retrovirus

heterozygous: A genetic state in which the maternal and paternal copies of a gene, or the arrangements within the maternal and paternal copies of a chromosome, are different.

hominid: Any species classified in the family Hominidae, which includes humans and great apes, as well as their extinct relatives dating back to the common ancestor of humans and great apes.

hominin: Any species within the hominin clade, which includes humans and their extinct relatives that lived since their ancestral lineage diverged from the panin clade, the lineage leading to the common chimpanzee and bonobo.

Homo: The genus that includes modern humans and their extinct relatives dating to the emergence of *Homo* from australopithecines approximately two million years ago.

Homo sapiens: The Latin binomial species name of modern humans.

homozygous: A genetic state in which the maternal and paternal copies of a gene, or the arrangements within the maternal and paternal copies of a chromosome, are the same.

host: An individual who has been infected by an organism or virus.

human endogenous retrovirus (HERV): A retrovirus that infected a common ancestor of all humans and integrated itself into the germline of that ancestor so that it now is present in the genome of all modern humans.

hybrid: An individual that is the offspring of two genetically distant parents, sometimes parents that are of different but related species.

inbreeding: Mating between two individuals who share at least one recent common ancestor, usually within a few generations.

intelligent design: A movement promoting the belief that an intelligent being designed the universe, and especially life, excluding evolution, or limiting its role, as the mechanism for this design.

interspecific hybrid offspring: Offspring from a mating between individuals of different species, such as a horse and donkey mating to produce a mule.

interspecific mating: Mating between individuals of different species.

intron: A segment of DNA transcribed into RNA that ultimately is removed from the final version of RNA encoded by a gene.

inversion: *See* chromosome inversion

lineage: The ancestral line of individuals through whom DNA has been inherited from parent to offspring through multiple generations.

macrohaplogroup: An ancestral haplogroup from which a large number of haplogroups evolved. In humans, the M, N, and R mitochondrial haplogroups are considered to be macrohaplogroups.

mass selection: An ancient form of artificial selection in which people save the best plants or animals as breeding stock for the next generation and repeat the process with each generation.

maternal lineage: A lineage of genetic inheritance proceeding through multiple generations purely from mother to daughter, typified by the maternal inheritance of mitochondrial DNA.

microorganism: An organism that is microscopic in size.

mitochondrial DNA: DNA contained in the mitochondrion, which is inherited by all individuals exclusively from their mothers.

mitochondrial Eve: The most recent common ancestor for mitochondrial

DNA in humans. She lived in ancient Africa, and all living humans inherited their mitochondrial DNA from her.

modern synthesis: The synthesis in the early twentieth century of Mendelian genetics with Darwinian natural selection.

most recent common ancestor (MRCA): The most recent individual who carried a particular variant in DNA inherited by all individuals in a group of related individuals, a species, or a group of related species.

mutation: An inherited alteration in the sequence of DNA.

mutational decay: The accumulation of mutations in DNA, usually in DNA not subject to the effects of natural selection, such as pseudogenes.

naturalist: A person who studies nature.

natural selection: The preservation of beneficial inherited variations, and the elimination of detrimental inherited variations, among individuals in nature, resulting in gradual genetic change within a species and ultimately the evolution of new species.

negative purifying selection: A form of natural selection that results in the elimination of detrimental mutations, preserving the original DNA sequence in a population.

neo-Darwinism: The concept of Darwinian natural selection as informed by Mendelian genetics, which emerged from the modern synthesis.

paleontologist: A scientist who studies fossils.

paternal lineage: A lineage of genetic inheritance proceeding through multiple generations purely from father to son, typified by the paternal inheritance of the Y chromosome.

pathogen: An organism, usually a microorganism or virus, that infects an individual and is capable of causing disease.

poly(A) tail: A segment of adenine (A) bases, which is added to the end of an RNA molecule after it is transcribed from a gene.

positive selection: The preservation of beneficial mutations in DNA through natural selection.

primate: A group of related mammalian species that includes humans, apes, monkeys, lemurs, galagos, marmosets, tarsiers, and bush babies.

protein: A chain of amino acids encoded by DNA in a gene and translated from RNA, folded into a functional unit.

pseudogene: A copy of a gene or portion of a gene that does not function as a gene.

quadruped: An animal that has four legs (literally four feet) and habitually walks on all four.

recessive variant: A variation in DNA that is expressed in the outward appearance of an organism only when present on both the maternal and paternal copies of chromosomes.

retroelement: A segment of DNA in the genome of an individual or species that originated from reverse transcription of RNA into DNA.

retrogene: A retroelement that functions as a gene.

retropseudogene: A pseudogene derived from reverse transcription of an RNA encoded by a gene into DNA.

retrotransposition: A process that begins when an RNA molecule is transcribed from DNA and ends when the RNA is reverse transcribed into DNA and that DNA is inserted elsewhere in the genome.

retrovirus: A virus with RNA as its genetic material. After the retrovirus infects a cell, its RNA is reverse transcribed into DNA, and the retroviral DNA is then inserted into the DNA of the infected cell.

RNA: Ribonucleic acid, a linear molecule composed of nucleotides, which contain the bases U, C, A, and G, and is usually transcribed from genes in DNA.

robust: A group of ancient hominins, characterized by a crest on the top of the cranium, that lived from approximately three to one million years ago. Robusts are descended from australopithecines and are sometimes classified as late australopithecines.

sedimentary rocks: Rocks that formed from the compression of sediments that collected over long periods of geologic time. Most fossils are found in sedimentary rocks.

selective sweep: A form of intense positive selection that favors a particular variant so that the variant, and other variants near it in the DNA, rapidly displace other variants until they are present in every individual in the species.

selectively neutral mutations: Mutations that are neither favored nor disfavored through natural selection.

special creation: The idea that God created all kinds of life separately and distinctly in the same form they currently have.

species: A group of individuals that are able to mate with one another and produce fully fertile offspring and are reproductively isolated from other such groups.

superhaplogroup: A large collection of diverse haplogroups that trace their origin to the same ancestral haplogroup. In humans, the mitochondrial L haplogroups constitute a superhaplogroup and contain more genetic diversity than all other mitochondrial haplogroups combined.

telomerase: An enzyme that restores DNA eroded on the ends of a linear DNA molecule after DNA replication.

telomere: The end of a linear DNA molecule in a chromosome. Telomeres contain a particular six-base-pair DNA sequence repeated in tandem, in humans: TTAGGG.

tetrapod: An organism with four appendages (literally four feet), usually used for locomotion.

transcription: The transfer of genetic information from DNA to RNA through the synthesis of an RNA molecule from a gene in DNA.

transitional form: A species with characteristics that are transitional between the characteristics that typify major groups of organisms.

translation: The transfer of genetic information from RNA to protein through the synthesis of a protein from an RNA molecule encoded by a gene in DNA.

translocation: *See* chromosome translocation

unbalanced cell: A cell that is missing chromosomes or segments of chromosomes, or has extra chromosomes or segments of chromosomes.

variety: A group of organisms that has distinctive inherited characteristics when compared with other such groups within the same species.

vestige: A remnant of an anatomical structure whose original function is reduced or absent.

virus: An infectious particle consisting of DNA or RNA surrounded by a protein coat.

Y-chromosome Adam: The most recent common ancestor for Y-chromosome DNA in humans. He lived in ancient Africa, and all living human males inherited their Y-chromosome DNA from him.

Y-chromosome DNA: The DNA molecule in the Y chromosome, which is inherited in a purely paternal fashion, exclusively from father to son.

NOTES

PROLOGUE

1. F. Darwin, ed., *The Life and Letters of Charles Darwin*, vol. 1 (New York: D. Appleton and Company, 1887), p. 473.

2. Ibid., p. 453.

3. Ibid., p. 467.

4. C. R. Darwin, *On the Origin of Species by Means of Natural Selection, or the Preservation of Favoured Races in the Struggle for Life* (London: John Murray, 1859), p. 488.

CHAPTER 1. WHAT IS EVOLUTION?

1. T. Dobzhansky, "Nothing in Biology Makes Sense Except in the Light of Evolution," *American Biology Teacher* 35 (1973): 125–29.

2. J. Ray, *The Wisdom of God Manifested in the Works of Creation* (London: R. Harbin at the Prince's-Arms in St Paul's Church Yard, 1717), preface.

3. C. R. Darwin, *Narrative of the Surveying Voyages of His Majesty's Ships* Adventure *and* Beagle *between the Years 1826 and 1836, Describing Their Examination of the Southern Shores of South America, and the* Beagle's *Circumnavigation of the Globe. Journal and Remarks. 1832–1836*, vol. 3 (London: Henry Colburn, 1839), p. 462.

4. C. R. Darwin, *On the Origin of Species by Means of Natural Selection, or the Preservation of Favoured Races in the Struggle for Life* (London: John Murray, 1859), p. 406.

5. Ibid., p. 61.

6. G. Mendel, "Experiments in Plant Hybridization," in *Mendel's Principles of Heredity*, W. Bateson, trans. C. T. Druery (Cambridge: Cambridge University Press, 1913), p. 335.

7. D. J. Fairbanks and B. Rytting, "Mendelian Controversies: A Botanical and Historical Review," *American Journal of Botany* 88 (2001): 737–52. This article reviews the disparate views on the connection between Darwin and Mendel.

8. O. T. Avery, C. M. MacLeod, and M. McCarty, "Studies on the Chemical Nature of the Substance Inducing Transformation of the Pneumococcal Types," *Journal of Experimental Medicine* 79 (1944): 137–58; A. D. Hershey and M. Chase, "Independent Functions of Viral

Protein and Nucleic Acid in Growth of Bacteriophage," *Journal of General Physiology* 36 (1952): 39–56.

9. Chimpanzee Sequencing and Analysis Consortium, "Initial Sequence of the Chimpanzee Genome and Comparison with the Human Genome," *Nature* 437 (2005): 69–87.

10. International Human Genome Sequencing Consortium, "Initial Sequencing and Analysis of the Human Genome," *Nature* 409 (2001): 860–921.

11. International Human Genome Sequencing Consortium, "Finishing the Euchromatic Sequence of the Human Genome," *Nature* 431 (2004): 931–45.

12. Chimpanzee Sequencing and Analysis Consortium, "Initial Sequence of the Chimpanzee Genome."

13. Rhesus Macaque Genome Sequencing and Analysis Consortium, "Evolutionary and Biomedical Insights from the Rhesus Macaque Genome," *Science* 316 (2007): 222–34.

14. Gallup, "Evolution, Creationism, Intelligent Design," http://www.gallup.com/poll/21814/evolution-creationism-intelligent-design.aspx (accessed April 3, 2011).

15. For examples, see D. J. Fairbanks, *Relics of Eden: The Powerful Evidence of Evolution in Human DNA* (Amherst, NY: Prometheus Books, 2007); E. C. Scott, *Evolution vs. Creationism: An Introduction* (Westport, CT: Greenwood Press, 2004); K. W. Miller, *Finding Darwin's God: A Scientist's Search for Common Ground between God and Evolution* (New York: Cliff Street Books, 1999); and F. S. Collins, *The Language of God: A Scientist Presents Evidence for Belief* (New York: Free Press, 2006).

CHAPTER 2. EVIDENCE FROM OUR BODIES

1. E. Haeckel, *Kunstformen der Natur* (Leipzig: Bibliographisches Institut, 1904).

2. R. Dawkins, *The Blind Watchmaker* (New York: W. W. Norton, 1986), p. 36.

3. L. I. Held Jr., *Quirks of Human Anatomy: An Evo-Devo Look at the Human Body* (Cambridge: Cambridge University Press, 2009), pp. 115–23; R. Dawkins, *The Greatest Show on Earth: The Evidence for Evolution* (New York: Free Press, 2009), pp. 352–55; J. Coyne, *Why Evolution Is True* (New York: Viking, 2009), pp. 141–43.

4. C. R. Darwin, *On the Origin of Species by Means of Natural Selection, or the Preservation of Favoured Races in the Struggle for Life* (London: John Murray, 1859), p. 61.

5. C. R. Darwin, *On the Origin of Species by Means of Natural Selection, or the Preservation of Favoured Races in the Struggle for Life*, 6th ed. (London: John Murray, 1872), p. 143. When quoting from *The Origin of Species*, I prefer to use the first edition. However, Darwin clarified the language in this quoted passage to make his point more emphatic in subsequent editions, the sixth being the best, which I've quoted here.

6. Ibid., pp. 143–44.

7. J. A. Bar-Maor, K. M. Kesner, and J. K. Karftori, "Human Tails," *Journal of Bone and Joint Surgery* 62 (1980): 508–10.

8. C. Stanford, *Upright: The Evolutionary Key to Becoming Human* (Boston: Houghton Mifflin, 1980). This excellent book on the evolution of habitual upright bipedalism is a principal source for much of the information on bipedalism in this chapter.

9. T. M. Greiner, "The Morphology of the Gluteus Maximus during Human Evolution: Prerequisite or Consequence of the Upright Bipedal Posture?" *Human Evolution* 17 (2006): 79–94.

10. C. B. Cunningham et al., "The Influence of Foot Posture on the Cost of Transport in Humans," *Journal of Experimental Biology* 213 (March 1, 2010): 790–97.

11. G. J. Sawyer et al., *The Last Human: A Guide to Twenty-Two Species of Extinct Humans* (New Haven, CT: Yale University Press, 2007), p. 89.

12. A leading researcher in this area who has devoted much of his career to the relationship of breathing and locomotion with evolution is David Carrier at the University of Utah. Notably, he developed a concept now called Carrier's Constraint, which explains the evolution of breathing in the transition from reptiles to mammals, introduced in the following article: D. R. Carrier, "The Evolution of Locomotor Stamina in Tetrapods: Circumventing a Mechanical Constraint," *Paleobiology* 13 (1987): 326–41.

13. Stanford, *Upright*, p. 55.

14. A. R. Silvestri Jr. and I. Singh, "The Unresolved Problem of the Third Molar: Would People Be Better Off without It?" *Journal of the American Dental Association* 134: 450–55; G. Arora, N. Polavarapu, and J. F. McDonald, "Did Natural Selection for Increased Cognitive Ability in Humans Lead to an Elevated Risk of Cancer?" *Medical Hypotheses* 73 (2009): 453–56; B. J. Crespi and K. Summers, "Positive Selection in the Evolution of Cancer," *Biological Review of the Cambridge Philosophical Society* 81 (2006): 407–24.

CHAPTER 3. EVIDENCE FROM THE EARTH

1. N. Shubin, *Your Inner Fish* (New York: Pantheon Books, 2008).

2. L. Rook et al., "Oreopithecus Was a Bipedal Ape after All: Evidence from the Iliac Cancellous Architecture," *Proceedings of the National Academy of Sciences, USA* 96 (1999): 8795–99.

3. G. Suwa et al., "A New Species of Great Ape from the Late Miocene Epoch in Ethiopia," *Nature* 448 (2007): 921–24.

4. S. McBrearty and N. G. Jablonski, "First Fossil Chimpanzee," *Nature* 437 (2005): 105–108.

5. G. J. Sawyer et al., *The Last Human: A Guide to Twenty-Two Species of Extinct Humans* (New Haven, CT: Yale University Press, 2007).

6. R. Dawkins, *The Greatest Show on Earth: The Evidence for Evolution* (New York: Free Press, 2010), p. 199.

7. The summaries of different hominin species in this part of the chapter were derived principally from five sources: G. J. Sawyer et al., *The Last Human*; I. Tattersall and J. H. Schwartz, *Extinct Humans* (New York: Névraumont, 2000); the October 2, 2009, issue of the journal *Science* 326, no. 5949, which contains multiple articles describing the recent reconstructions and analyses of *Ardipithecus ramidus*, placing it in the context of other hominins; C. Zimmer, *Smithsonian Intimate Guide to Human Origins* (Toronto, ON, Canada: Madison Press Books, Smithsonian Books, and Collins, 2005); and the TalkOrigins Archive, "Fossil Hominids: The Evidence for Human Evolution," http://www.talkorigins.org/faqs/homs (accessed July 27, 2011).

8. Y. Haile-Selassie, G. Suwa, and T. D. White, "Late Miocene Teeth from Middle Awash, Ethiopia, and Early Hominid Dental Evolution," *Science* 303 (2004): 1503–1505.

9. T. D. White et al., "*Ardipithecus ramidus* and the Paleobiology of Early Hominids," *Science* 326 (2009): 75–86.

10. G. Suwa et al., "The *Ardipithecus ramidus* Skull and Its Implications for Hominid Origins," *Science* 326 (2009): 68, 68e1–e7.

11. C. O. Lovejoy et al., "Careful Climbing in the Miocene: The Forelimbs of *Ardipithecus ramidus* and Humans Are Primitive," *Science* 326 (2009): 70, 70e1–e8; C. O. Lovejoy et al., "Combining Prehension and Propulsion: The Foot of *Ardipithecus ramidus*," *Science* 326 (2009): 72, 72e1–e8.

12. C. O. Lovejoy et al., "The Pelvis and Femur of *Ardipithecus ramidus*: The Emergence of Upright Walking," *Science* 326 (2009): 71, 71e1–e6.

13. M. D. Leakey and R. L. Hay, "Pliocene Footprints in the Laetoli Beds at Laetoli, Northern Tanzania," *Nature* 278 (1979): 317–23.

14. D. A. Raichlen et al., "Laetoli Footprints Preserve Earliest Direct Evidence of Human-Like Bipedal Biomechanics," *PLoS ONE* 5 (2010): e9769.

15. C. V. Ward, W. H. Kimbel, and D. C. Johanson, "Complete Fourth Metatarsal and Arches in the Foot of *Australopithecus afarensis*," *Science* 331 (2011): 750–53.

16. J. de Heinzelin et al., "Environment and Behavior of 2.5-Million-Year-Old Bouri Hominids," *Science* 284 (1999): 625–29.

17. G. J. Sawyer et al., *The Last Human*, pp. 81, 109–10, 136; R. L. Susman, "Fossil Evidence for Early Hominid Tool Use," *Science* 265 (1994): 1570–73.

18. L. R. Berger et al., "*Australopithecus sediba*: A New Species of *Homo*-Like Australopith from South Africa," *Science* 328 (2010): 195–204.

19. G. J. Sawyer et al., *The Last Human*, p. 127; R. L. Susman, "Fossil Evidence for Early Hominid Tool Use."

20. S. W. Simpson et al., "A Female *Homo erectus* Pelvis from Gona, Ethiopia," *Science* 322 (2008): 1089–92.

21. For a recent review of the debate, see L. C. Aiello, "Five Years of *Homo floresiensis*," *American Journal of Physical Anthropology* 142 (2008): 167–79.

22. G. J. Sawyer et al., *The Last Human*, p. 215.

23. C. Finlayson et al., "Late Survival of Neanderthals at the Southernmost Extreme of Europe," *Nature* 443 (2006): 850–53.

24. J. Diamond, *Guns, Germs, and Steel* (New York: W. W. Norton, 1999).

25. R. E. Green et al., "A Draft Sequence of the Neandertal Genome," *Science* 328 (2010): 710–22.

26. D. Reich et al., "Genetic History of an Archaic Hominin Group from Denisova Cave in Siberia," *Nature* 468 (2010): 1053–60.

CHAPTER 4. EVIDENCE FROM GEOGRAPHY

1. M. Pei, *The Story of Latin and the Romance Languages* (New York: Harper & Row, 1976), pp. 160–61.

2. This English translation is my own, and I made it as literally close as possible to the original Italian. Others have translated this sonnet differently, usually sacrificing a literal translation to facilitate rhyming in sonnet form. My preference when translating poetry, however, is to abandon forced rhyming and preserve the meaning as much as possible. For those who are interested, here is the full sonnet in the original Italian, followed by my English translation. Notice how the final words of each verse in Michelangelo's original Italian version rhyme in sonnet form.

> *Veggio co' be' vostr'occhi un dolce lume*
> *che co' mie ciechi già veder non posso;*
> *porto co' vostri piedi un pondo addosso,*
> *che de' mie zoppi non è già costume.*
> *Volo con le vostr'ale senza piume;*
> *col vostro ingegno al ciel sempre son mosso;*
> *dal vostro arbitrio son pallido e rosso,*
> *freddo al sol, caldo alle più fredde brume.*
> *Nel voler vostro è sol la voglia mia,*
> *i miei pensier nel vostro cor si fanno,*
> *nel vostro fiato son le mie parole.*
> *Come luna da sé sol par ch'io sia,*
> *ché gli occhi nostri in ciel veder non sanno*
> *se non quel tanto che n'accende il sole.*

> Through your beautiful eyes I see a sweet light
> that my blind eyes cannot yet see;
> Upon your feet I carry a great burden
> that my lame feet cannot yet bear.

> I fly without feathers upon your wings;
> by your genius I am forever carried to the sky;
> By the power of your will I am pale and red,
> cold in the sun, yet warmed in the coldest mist.
> Your wants are my only desire,
> my thoughts form within your heart,
> my words upon your breath.
> I seem like the moon alone,
> as our eyes in the heavens see but cannot understand,
> until illuminated by the sun.

3. My sincere thanks to Constantin Caradja and Christa Albrecht-Crane for their assistance with the Romanian translation, and to Louse Illes for her assistance with the French translation. The other translations are my own.

4. F. Sanger et al., "Nucleotide Sequence of Bacteriophage φX174 DNA," *Nature* 265 (1977): 687–95.

5. Rarely in biology are rules absolute, and this is true for inheritance of mitochondrial DNA. There is a possibility that an exceptionally small amount of mitochondrial DNA may be inherited from the father. There are no documented cases of paternal inheritance of mitochondrial DNA in humans, and there is some evidence that the few paternal mitochondria in the sperm cell that enter the egg cell at fertilization are selectively destroyed. However, breeding experiments in mice specifically designed to enrich mitochondrial DNA inherited from the father over many generations result in an exceptionally small fraction of paternally transmitted DNA detected in a small number of individuals, indicating that in rare circumstances a very few paternal mitochondria may be transmitted. Because there is no situation in humans comparable to these experiments, there is currently no possibility of determining whether or not a small amount of paternal mitochondrial inheritance has influenced human evolutionary history. Technically, it is most correct to say that mitochondrial inheritance in mammals is overwhelmingly maternal, although for practical purposes we can consider it to be purely maternal.

6. There are rare exceptions. Some human females have an XY chromosome constitution, usually due to mutations, and some of those mutations are in the Y chromosome itself. Females who inherit a Y chromosome, however, are infertile and do not pass it on to offspring.

7. L. L. Cavalli-Sforza and F. Cavalli-Sforza, *The Great Human Diasporas: The History of Diversity and Evolution* (Reading, MA: Addison-Wesley, 1995), p. 106.

8. The Genographic Project's website can be accessed at http://www.genographic.org.

9. This tree is based on two comprehensive studies incorporating information from complete mitochondrial DNA sequencing as of 2009. The two studies are N. van Oven and M. Kayser, "Updated Comprehensive Phylogenetic Tree of Global Human Mitochondrial DNA Variation," *Human Mutation* 30 (2009): E386–94, and P. Soares et al., "Correcting for Purifying

Selection: An Improved Human Mitochondrial Molecular Clock," *American Journal of Human Genetics* 84 (2009): 740–59.

10. M. Ingman et al., "Mitochondrial Genome Variation and the Origin of Modern Humans," *Nature* 408 (2000): 708–13.

11. Van Oven and Kayser, "Updated Comprehensive Phylogenetic Tree."

12. Dating the origin of the mitochondrial Eve and mitochondrial haplogroups has been substantially refined in recent years as whole mitochondrial genome sequences have become available. Throughout this chapter, I have used the dates published by P. Soares et al., "Correcting for Purifying Selection." A more recent article has extended the date of the mitochondrial Eve to 200,000 years ago: K. A. Cyran and M. Kimmel, "Alternatives to the Wright-Fisher Model: The Robustness of Mitochondrial Eve Dating," *Theoretical Population Biology* 78 (2010): 165–72. Both 192,000 and 200,000 years ago are within the margins of error for both estimates. Soares et al. provide dates not only for the mitochondrial Eve but also for all mitochondrial haplogroups. I have used their dates for mitochondrial haplogroups throughout this chapter for consistency.

13. I. S. Zalmout et al., "New Oligocene Primate from Saudi Arabia and the Divergence of Apes and Old World Monkeys," *Nature* 466 (2010): 360–64.

14. M. K. Gonder et al., "Whole-mtDNA Genome Sequence Analysis of Ancient African Lineages," *Molecular Biology and Evolution* 24 (2007): 757–68.

15. I. S. Castañeda et al., "Wet Phases in the Sahara/Sahel Region and Human Migration Patterns in North Africa," *Proceedings of the National Academy of Sciences, USA* 106 (2009): 20159–63.

16. S. Wells, *Deep Ancestry: Inside the Genographic Project* (Washington, DC: National Geographic Society, 2006), p. 183.

17. J. Friedlaender et al., "Expanding Southwest Pacific Mitochondrial Haplogroups P and Q," *Molecular Biology and Evolution* 22 (2005): 1506–17.

18. P. Soares et al., "Climate Change and Postglacial Human Dispersals in Southeast Asia," *Molecular Biology and Evolution* 25 (2008): 1209–18.

19. S. Wells, *Deep Ancestry*.

20. M. V. Derenko et al., "The Presence of Mitochondrial Haplogroup X in Altaians from South Siberia," *American Journal of Human Genetics* 69 (2001): 237–41.

21. U. A. Perego et al., "Distinctive Paleo-Indian Migration Routes from Beringia Marked by Two Rare mtDNA Haplogroups," *Current Biology* 13 (2009): 1–8; N. V. Volodko et al., "Mitochondrial Genome Diversity in Arctic Siberians, with Particular Reference to the Evolutionary History of Beringia and Pleistocenic Peopling of the Americas," *American Journal of Human Genetics* 82 (2008): 1084–1100; R. S. Malhi and D. G. Smith, "Haplogroup X Confirmed in Prehistoric North America," *American Journal of Physical Anthropology* 119 (2002): 84–86.

22. P. de Knijff, "Messages through Bottlenecks: On the Combined Use of Slow and Fast

Evolving Polymorphic Markers on the Human Y Chromosome," *American Journal of Human Genetics* 67 (2000): 1055–61.

23. T. M. Karafet et al., "New Binary Polymorphisms Reshape and Increase Resolution of the Human Y Chromosomal Haplogroup Tree," *Genome Research* 18 (2008): 830–38.

24. Ibid.

25. Examples include a colossal marble statue in Florence by Giambologna and two large paintings by Nicolas Poussin, one in the Louvre in Paris and the other in the Metropolitan Museum of Art in New York City.

26. O. Semino et al., "Ethiopians and Khoisan Share the Deepest Clades of the Human Y-Chromosome Phylogeny," *American Journal of Human Genetics* 70 (2002): 265–68.

27. Ibid.

28. T. Zerjal et al., "The Genetic Legacy of the Mongols," *American Journal of Human Genetics* 72 (2003): 717–21.

29. L. A. Rodriguez-Delfin, V. E. Rubin-de-Celis, and M. A. Zago, "Genetic Diversity in an Andean Population from Peru and Regional Migration Patterns of Amerindians in South America: Data from Y Chromosome and Mitochondrial DNA." *Human Heredity* 51 (2001): 97–106.

30. M. F. Hammer et al., "Population Structure of Y Chromosome SNP Haplogroups in the United States and Forensic Implications for Constructing Y Chromosome STR Databases," *Forensic Science International* 164 (2006): 45–55.

31. S. L. Zegura et al., "High-Resolution SNPs and Microsatellite Haplotypes Point to a Single, Recent Entry of Native American Y Chromosomes into the Americas," *Molecular Biology and Evolution* 21 (2004): 164–75.

32. Ibid., p. 171.

CHAPTER 5. EVIDENCE FROM OUR GENOME: EONS OF SHUFFLING AND REARRANGING

1. M. E. Drets and M. W. Shaw, "Specific Banding Patterns of Human Chromosomes," *Proceedings of the National Academy of Sciences, USA* 68 (1971): 2073–77.

2. Human chromosomes were initially designated as 1 through 22 on the basis of decreasing size. However, as methods improved, geneticists discovered that chromosome 21 is slightly smaller than chromosome 22. However, because of the large number of publications with the name "chromosome 21" specifically applied to this smallest chromosome, largely because of its role in Down syndrome, the original numbering system has been retained.

3. J. J. Yunis, J. R. Sawyer, and K. Dunham, "The Striking Resemblance of High-Resolution G-Banded Chromosomes of Man and Chimpanzee," *Science* 208 (1980): 1145–48.

4. International Human Genome Sequencing Consortium, "Initial Sequencing and Analysis of the Human Genome," *Nature* 409 (2001): 860–921.

5. Chimpanzee Sequencing and Analysis Consortium, "Initial Sequence of the Chimpanzee Genome and Comparison with the Human Genome," *Nature* 437 (2005): 69–87.

6. C. W. Greider and E. H. Blackburn, "Identification of a Specific Telomere Terminal Transferase Activity in *Tetrahymena* Extracts," *Cell* 43 (1985): 405–13.

7. A scientific literature search performed on March 29, 2011, found 10,044 scientific articles published on telomerase since Greider and Blackburn's 1985 article.

8. V. Goidts et al., "Segmental Duplication Associated with the Human-Specific Inversion of Chromosome 18: A Further Example of the Impact of Segmental Duplications on Karyotype and Genome Evolution in Primates," *Human Genetics* 115 (2004): 116–22.

9. G. Collodel et al., "TEM, FISH and Molecular Studies in Infertile Men with Pericentric Inversion of Chromosome 9," *Andrologia* 38 (2006): 122–27; M. Srebniak et al., "Subfertile Couple with inv(2),inv(9) and 16qh+," *Journal of Applied Genetics* 45 (2004): 477–79; I. P. Davalos et al., "Inv(9)(p24q13) in Three Sterile Brothers," *Annals of Genetics* 43 (2000): 51–54; I. Sasagawa et al., "Pericentric Inversion of Chromosome 9 in Infertile Men," *International Urology and Nephrology* 30 (1998): 203–207.

10. G. Montefalcone et al., "Centromere Repositioning," *Genome Research* 9 (1999): 1184–88.

11. H. Kehrer-Sawatzki et al., "Molecular Characterization of the Pericentric Inversion of Chimpanzee Chromosome 11 Homologous to Human Chromosome 9," *Genomics* 85 (2005): 542–50.

12. R. Stanyon et al., "Primate Chromosome Evolution: Ancestral Karyotypes, Marker Order and Neocentromeres," *Chromosome Research* 16 (2008):17–39.

13. J. J. Ely et al., "Technical Note: Chromosomal and mtDNA Analysis of Oliver," *American Journal of Physical Anthropology* 105 (1998): 395–403.

14. H. Klaatsch, *The Evolution and Progress of Mankind*, trans. J. McCabe (New York: F. A. Stokes, 1923).

15. A. Etkind, "Beyond Eugenics: The Forgotten Scandal of Hybridizing Humans and Apes," *Studies in History and Philosophy of Biological and Biomedical Sciences* 39 (2008): 205–210.

16. My description of Ivanov's experiments is drawn largely from Etkind, "Beyond Eugenics."

17. Chimpanzee Sequencing and Analysis Consortium, "Initial Sequence of the Chimpanzee Genome."

18. J. M. Szamalek et al., "The Chimpanzee-Specific Pericentric Inversions that Distinguish Humans and Chimpanzees Have Identical Breakpoints in *Pan troglodytes* and *Pan paniscus*," *Genomics* 87 (2006): 39–45.

19. R. E. Green et al., "A Draft Sequence of the Neandertal Genome," *Science* 328 (2010): 710–12; D. Reich et al., "Genetic History of an Archaic Hominin Group from Denisova Cave in Siberia," *Nature* 468 (2010): 1053–60.

CHAPTER 6. MORE EVIDENCE FROM OUR GENOME: THE GENES THAT MAKE US HUMAN

1. V. A. McKusick, *Mendelian Inheritance in Man: A Catalog of Human Genes and Genetic Disorders*, vol. 3, 12th ed. (Baltimore: Johns Hopkins University Press, 1998): 2202.

2. International Human Genome Sequencing Consortium, "Finishing the Euchromatic Sequence of the Human Genome," *Nature* 431 (2004): 931–45.

3. For examples, see M. H. Behe, *The Edge of Evolution: The Search for the Limits of Darwinism* (New York: Free Press, 2007); J. C. Sanford, *Genetic Entropy and the Mystery of the Genome*, 3rd ed. (Waterloo, NY: FMS, 2008).

4. C. S. Lai et al., "A Forkhead-Domain Gene Is Mutated in a Severe Speech and Language Disorder," *Nature* 413 (2001): 519–23.

5. J. Zhang, D. M. Webb, and O. Podlaha, "Accelerated Protein Evolution and Origins of Human-Specific Features: *FOXP2* as an Example," *Genetics* 162 (2002): 1825–35.

6. J. Kraus et al., "The Derived *FOXP2* Variant of Modern Humans Was Shared with Neandertals," *Current Biology* 17 (2007): 1–5; G. Coop et al., "The Timing of Selection at the Human *FOXP2* Gene," *Molecular Biology and Evolution* 25 (2008): 1257–59; S. E. Ptak et al., "Linkage Disequilibrium Extends across Putative Selective Sites in *FOXP2*," *Molecular Biology and Evolution* 26 (2009): 2181–84.

7. Zhang, Webb, and Podlaha, "Accelerated Protein Evolution."

8. S. M. Rich et al., "The Origin of Malignant Malaria," *Proceedings of the National Academy of Sciences, USA* 106 (2009): 14902–907.

9. Ibid.

10. H. H. Chou et al., "Inactivation of CMP-N-acetylneuraminic Acid Hydroxylase Occurred Prior to Brain Expansion during Human Evolution," *Proceedings of the National Academy of Sciences, USA* 99 (2002): 11736–41.

11. World Health Organization, *World Malaria Report 2008* (Geneva, Switzerland: WHO Press, 2008).

12. A. V. S. Hill et al., "Polynesian Origins and Affinities: Globin Gene Variants in Eastern Polynesia," *American Journal of Human Genetics* 40 (1987): 453–63.

13. D. Torrents et al., "A Genome-Wide Survey of Human Pseudogenes," *Genome Research* 13 (2003): 2559–67.

14. R. Kurzrock et al., "Philadelphia Chromosome-Positive Leukemias: From Basic Mechanisms to Molecular Therapeutics," *Annals of Internal Medicine* 138 (2003): 819–30.

15. P. Rogalla et al., "Back to the Roots of a New Exon—The Molecular Archaeology of a SP100 Splice Variant," *Genomics* 63 (2000): 117–22.

16. D. G. Knowles and A. McLysaght, "Recent De Novo Origin of Human Protein-Coding Genes," *Genome Research* 19 (2009): 1752–59.

17. Chimpanzee Sequencing and Analysis Consortium, "Initial Sequence of the

Chimpanzee Genome and Comparison with the Human Genome," *Nature* 437 (2007): 69–87.

18. Ibid.

19. Rhesus Macaque Genome Sequencing and Analysis Consortium, "Evolutionary and Biomedical Insights from the Rhesus Macaque Genome," *Science* 316 (2007): 222–34.

20. Chimpanzee Sequencing and Analysis Consortium, "Initial Sequence of the Chimpanzee Genome."

21. S. Dorus et al., "Accelerated Evolution of Nervous System Genes in the Origin of *Homo sapiens*," *Cell* 119 (2004): 1027–40.

22. C. P. Bird, "Fast-Evolving Noncoding Sequences in the Human Genome," *Genome Biology* 8 (2007): R118.

Chapter 7. Even More Evidence from Our Genome: Invaders and Freeloaders by the Millions

1. International Human Genome Sequencing Consortium, "Initial Sequencing and Analysis of the Human Genome," *Nature* 409 (2001): 860–921.

2. M. Dewannieux et al., "Identification of an Infectious Progenitor for the Multiple-Copy HERV-K Human Endogenous Retroelements," *Genome Research* 16 (2006): 1548–56; Y. N. Lee and P. D. Bieniasz, "Reconstitution of an Infectious Human Endogenous Retrovirus," *PLoS Pathology* 3 (2007): e10.

3. International Human Genome Sequencing Consortium, "Initial Sequencing and Analysis of the Human Genome."

4. Ibid.

5. D. Torrents et al., "A Genome-Wide Survey of Human Pseudogenes." *Genome Research* 13 (2003): 2559–67.

6. Y. J. Liu et al., "Comprehensive Analysis of the Pseudogenes of Glycolytic Enzymes in Vertebrates: The Anomalously High Number of *GAPDH* Pseudogenes Highlights a Recent Burst of Retrotrans-Positional Activity," *BMC Genomics* 10 (2009): 480.

7. Z. Zhang et al., "Millions of Years of Evolution Preserved: A Comprehensive Catalog of the Processed Pseudogenes in the Human Genome," *Genome Research* 13 (2003): 2541–58.

8. J. Young, J. Ménétrey, and B. Goud, "*RAB6C* Is a Retrogene That Encodes a Centrosomal Protein Involved in Cell Cycle Progression," *Journal of Molecular Biology* 397 (2010): 69–88; X. F. Liu et al., "A Primate-Specific POTE-Actin Fusion Protein Plays a Role in Apoptosis," *Apoptosis* 14 (2009): 1237–44; C. R. Jeter et al., "Functional Evidence That the Self-Renewal Gene *NANOG* Regulates Human Tumor Development," *Stem Cells* 27 (2009): 993–1005; J. Zhang et al., "*NANOGP8* Is a Retrogene Expressed in Cancers," *FEBS Journal* 273 (2006): 1723–30.

9. R. J. Britten, "Coding Sequences of Functioning Human Genes Derived Entirely from Mobile Element Sequences," *Proceedings of the National Academy of Sciences, USA* 101 (2004): 16825–30.

10. International Human Genome Sequencing Consortium, "Initial Sequencing and Analysis of the Human Genome."

11. Y. Quentin, "Origin of the *Alu* Family: A Family of Alu-Like Monomers Gave Birth to the Left and the Right Arms of the *Alu* Elements," *Nucleic Acids Research* 20 (1992): 3397–401; J. Jurka and E. Zuckerkandl, "Free Left Arms as Precursor Molecules in the Evolution of *Alu* Sequences," *Journal of Molecular Evolution* 33 (1991): 49–56; Y. Quentin, "Fusion of a Free Left *Alu* Monomer and a Free Right *Alu* Monomer at the Origin of the *Alu* Family in the Primate Genomes," *Nucleic Acids Research* 20 (1992): 487–93.

12. Rhesus Macaque Genome Sequencing and Analysis Consortium, "Evolutionary and Biomedical Insights from the Rhesus Macaque Genome," *Science* 316 (2007): 222–34.

13. I. Sawada et al., "Evolution of *Alu* Family Repeats since the Divergence of Human and Chimpanzee," *Journal of Molecular Evolution* 22 (1985): 316–22.

14. H. A. F. Booth and P. W. H. Holland, "Eleven Daughters of NANOG," *Genomics* 84 (2004): 229–38.

15. D. J. Fairbanks and P. J. Maughan, "Evolution of the *NANOG* Pseudogene Family in the Human and Chimpanzee Genomes," *BMC Evolutionary Biology* 6 (2006): 12.

16. C. R. Jeter et al., "Functional Evidence That the Self-Renewal Gene *NANOG* Regulates Human Tumor Development"; J. Zhang et al., "*NANOGP8* Is a Retrogene Expressed in Cancers."

CHAPTER 8. EVOLUTION AND OUR HEALTH

1. United States Department of Health and Human Services, Centers for Disease Control and Prevention, "Typhoid Fever," http://www.cdc.gov/nczved/divisions/dfbmd/diseases/typhoid _fever (accessed July 27, 2011).

2. P. M. Sharp and B. H. Hahn, "The Evolution of HIV-1 and the Origin of AIDS," *Philosophical Transactions of the Royal Society B* 365 (2010): 2487–94. This is an excellent review article. I used it as the principal source for most of the information I've presented here on HIV evolution.

3. N. Vidal et al., "Unprecedented Degree of Human Immunodeficiency Virus Type 1 (HIV-1) Group M Genetic Diversity in the Democratic Republic of Congo Suggests That the HIV-1 Pandemic Originated in Central Africa," *Journal of Virology* 74 (2000): 10498–507.

4. D. M. Mwaengo and F. J. Novembre, "Molecular Cloning and Characterization of Viruses Isolated from Chimpanzees with Pathogenic Human Immunodeficiency Virus Infections," *Journal of Virology* 72 (1998): 8976–87.

5. J. C. Plantier et al., "A New Human Immunodeficiency Virus Derived from Gorillas," *Nature Medicine* 15 (2009): 871–72.

6. F. Damond et al., "Identification of a Highly Divergent HIV Type 2 and Proposal for a Change in HIV Type 2 Classification," *AIDS Research and Human Retroviruses* 20 (2004): 666–72.

7. K. E. Garrison et al., "T Cell Responses to Human Endogenous Retroviruses in HIV-1 Infection," *PLoS Pathology* 3 (2007): e165.

8. T. R. Frieden et al., "The Emergence of Drug-Resistant Tuberculosis in New York City," *New England Journal of Medicine* 328 (1993): 521–26.

9. World Health Organization, *Global Tuberculosis Control: Epidemiology, Strategy, Financing* (Geneva, Switzerland: WHO; WHO/HTM/TB/2009.411, 2009).

10. R. Loddenkemper and B. Hauer, "Drug-Resistant Tuberculosis: A Worldwide Epidemic Poses a New Challenge," *Deutches Ärzteblatt International* 107 (2010): 10–19.

11. Y. Inai, Y. Ohta, and M. Nishikimi, "The Whole Structure of the Human Nonfunctional L-gulono-gamma-lactone Oxidase Gene—the Gene Responsible for Scurvy—and the Evolution of Repetitive Sequences Thereon," *Journal of Nutritional Science and Vitaminology* 49 (2003): 315–19.

12. A. Schuchat, B. P. Bell, and S. C. Redd, "The Science behind Preparing and Responding to Pandemic Influenza: The Lessons and Limits of Science," *Clinical Infectious Disease* 52 (2011): S8–12.

13. V. A. Belyi et al., "The Origins and Evolution of the p53 Family of Genes," *Cold Spring Harbor Perspectives in Biology* 2 (2010): a001198.

14. R. Rutkowski, K. Hofmann, and A. Gartner, "Phylogeny and Function of the Invertebrate p53 Superfamily," *Cold Spring Harbor Perspectives in Biology* 2 (2010): a001131.

15. V. A. Belyi et al., "Origins and Evolution of the p53 Family of Genes."

16. X. S. Puente et al., "Comparative Analysis of Cancer Genes in the Human and Chimpanzee Genomes," *BMC Genomics* 7 (2006): 15.

17. C. R. Jeter et al., "Functional Evidence that the Self-Renewal Gene *NANOG* Regulates Human Tumor Development"; J. Zhang et al., "*NANOGP8* Is a Retrogene Expressed in Cancers," *FEBS Journal* 273 (2006): 1723–30.

18. X. S. Puente et al., "Comparative Analysis of Cancer Genes in the Human and Chimpanzee Genomes."

19. M. C. Marin et al., "A Common Polymorphism Acts as an Intragenic Modifier of Mutant p53 Behaviour," *Nature Genetics* 25 (2000): 47–54; P. Dumont et al., "The Codon 72 Polymorphic Variants of p53 Have Markedly Different Apoptotic Potential," *Nature Genetics* 33 (2003): 357–65; J. S. Jones et al., "*p53* Polymorphism and Age of Onset of Hereditary Nonpolyposis Colorectal Cancer in a Caucasian Population," *Clinical Cancer Research* 10 (2004): 5845–49; S. Krüger et al., "The *p53* Codon 72 Variation Is Associated with the Age of Onset of Hereditary Non-Polyposis Colorectal Cancer (HNPCC)," *Journal of Medical Genetics* 42 (2005):

769–73; D. D. Ørsted, "Tumor Suppressor p53 Arg72Pro Polymorphism and Longevity, Cancer Survival, and Risk of Cancer in the General Population," *Journal of Experimental Medicine* 204 (2007): 1295–1301; S. El Hallani et al., "*TP53* Codon 72 Polymorphism Is Associated with Age at Onset of Glioblastoma," *Neurology* 72 (2009): 332–36.

20. G. A. Huttley et al., "Adaptive Evolution of the Tumour Suppressor *BRCA1* in Humans and Chimpanzees: Australian Breast Cancer Family Study," *Nature Genetics* 25 (2000): 410–13; M. A. Fleming et al., "Understanding Missense Mutations in the *BRCA1* Gene: An Evolutionary Approach," *Proceedings of the National Academy of Sciences, USA* 100 (2003): 1151–56; A. Pavlicek et al., "Evolution of the Tumor Suppressor *BRCA1* in Primates: Implications for Cancer Predisposition," *Human Molecular Genetics* 13 (2004): 2737–51.

21. R. Cook-Deegan et al., "Impact of Gene Patents and Licensing Practices on Access to Genetic Testing for Inherited Susceptibility to Cancer: Comparing Breast and Ovarian Cancers with Colon Cancers," *Genetics in Medicine* 12 (2010): S15–38.

22. L. F. Grochola et al., "Recent Natural Selection Identifies a Genetic Variant in a Regulatory Subunit of Protein Phosphatase 2A that Associates with Altered Cancer Risk and Survival," *Clinical Cancer Research* 15 (2009): 6301–308.

23. S. W. Brouilette et al., "Telomere Length, Risk of Coronary Heart Disease, and Statin Treatment in the West of Scotland Primary Prevention Study: A Nested Case-Control Study," *Lancet* 369 (2007): 107–14.

24. Examples include the movies *Gattaca* (1997) and *Code 46* (2003).

25. National Human Genome Research Institute, "Genetic Information Nondiscrimination Act (GINA) of 2008," http://www.genome.gov/24519851 (accessed July 26, 2011).

26. For those who are curious, I write this as I'm returning from a trip through Germany, Poland, Slovakia, Hungary, Austria, the Czech Republic, Switzerland, France, and Belgium. The official languages of these countries include German, Polish, Slovakian, Hungarian, Czech, French, Italian, and Flemish, not to mention the countless other languages we heard as we encountered people from all over the world who, like us, were traveling through these countries.

27. Sharp and Hahn, "The Evolution of HIV-1," p. 2942.

CHAPTER 9. EVOLUTION AND OUR FOOD

1. I've chosen to use "Ancient Pueblo People" instead of "Anasazi," which is a Navajo word that can be translated as "ancient enemy," and is considered offensive by some descendants of these people who currently live in the region.

2. K. E. Filchak, J. B. Roethele, and J. L. Feder, "Natural Selection and Sympatric Divergence in the Apple Maggot *Rhagoletis pomonella*," *Nature* 407 (2000): 739–42.

3. P. J. Maughan et al., "Characterization of Salt Overly Sensitive (*SOS1*) Gene Homoeologs in Quinoa (*Chenopodium quinoa* Willd)," *Genome* (2009): 647–57; M. R. B. Balzotti

et al., "Expression and Evolutionary Relationships of the *Chenopodium* quinoa 11S Seed Storage Protein Gene," *International Journal of Plant Science* 169 (2008): 281–91.

4. A. Franklin et al., *Ending the Mendel-Fisher Controversy* (Pittsburgh: University of Pittsburgh Press, 2007); D. J. Fairbanks and B. Rytting, "Mendelian Controversies: A Botanical and Historical Review," *American Journal of Botany* 88 (2001): 737–52.

5. United Nations Population Division, Department of Economic and Social Affairs, United Nations Secretariat. *The World at Six Billion*, http://www.un.org/esa/population/publications/sixbillion/sixbillion.htm (accessed October 26, 2010).

6. G. P. Nabhan, *Where Our Food Comes From: Retracing Nikolay Vavilov's Quest to End Famine* (Washington, DC: Island Press, 2009), p. 10.

7. Ibid., pp. 11–12.

8. I learned of these tributes personally in 1993 in a lengthy and unforgettable discussion with Anna Matalová, director of the Mendelianum Museum in Brno, Czech Republic. I am grateful to her for the time she generously gave me and for her passion for her work.

9. H. J. Dubin and J. P. Brennan, *Combating Stem and Leaf Rust of Wheat: Historical Perspective, Impacts, and Lessons Learned* (Washington, DC: International Food Policy Research Institute, 2009), available for free download at http://www.ifpri.org/publication/combating-stem-and-leaf-rust-wheat (accessed November 8, 2010).

10. P. N. Njau et al., "Identification and Evaluation of Sources of Resistance to Stem Rust Race Ug99 in Wheat," *Plant Disease* 94 (2010): 413–19.

11. Y. Jin et al., "Detection of Virulence to Resistance Gene *Sr24* within Race TTKS of *Puccinia graminis* f. sp. *tritici*," *Plant Disease* 92 (2008): 923–26.

12. For the latest information on Ug99 research, see http://www.globalrust.org and http://www.ars.usda.gov/Main/docs.htm?docid=14649 (both accessed April 12, 2011).

13. G. Frisvold, R. Tronstad, and J. Mortensen, "Effects of *Bt* Cotton Adoption: Regional Differences in Producer Costs and Returns," Final Research Project Report Cotton Incorporated, January 1999, University of Arizona, Department of Agricultural and Resource Economics, http://www.bt.ucsd.edu/bt_cotton.html (accessed November 21, 2010).

14. J. J. Duan et al., "A Meta-Analysis of Effects of *Bt* Crops on Honey Bees (Hymenoptera: Apidae)," *PLoS ONE* 3 (2008): e1415.

15. B. E. Tabashnik et al., "Insect Resistance to *Bt* Crops: Evidence versus Theory," *Nature Biotechnology* 26 (2008): 199–202.

16. S. B. Powles, "Evolved Glyphosate-Resistant Weeds around the World: Lessons to be Learnt," *Pest Management Science* 64 (2008): 360–65; M. D. K. Owen and I. A. Zelaya, "Herbicide-Resistant Crops and Weed Resistance to Herbicides," *Pest Management Science* 61 (2005): 301–311; I. M. Heap, "The Occurrence of Herbicide-Resistant Weeds Worldwide," *Pesticide Science* 51 (1997): 235–53.

CHAPTER 10. EVOLUTION AND OUR ENVIRONMENT

1. Apuane Geopark, "Inventory Card of the Geosites in the Apuan Alps Regional Park. No. 247: Fossils in Marble Banks in Foce di Pianza," http://www.parcapuane.toscana.it/apuane_geopark/ENGLISH%20VERSION/apuanegeopark_geosites.html (accessed December 16, 2010).

2. United Nations Population Division, *World Population Prospects: The 2000 Revision*, vol. 3 (New York: United Nations, 2000), p. 155.

3. Organisation for Economic Co-operation and Development, "OECD-FAO Agricultural Outlook 2010–2019," http://www.oecd.org/dataoecd/15/37/45599621.pdf (accessed December 31, 2010).

4. M. E. Conti and G. Cecchettib, "Biological Monitoring: Lichens as Bioindicators of Air Pollution Assessment—A Review," *Environmental Pollution* 114 (2001): 471–92.

5. J. Wells, *Icons of Evolution: Why Much of What We Teach about Evolution Is Wrong* (Washington, DC: Regnery, 2002); J. Hooper, *Of Moths and Men* (New York: W. W. Norton, 2002).

6. M. E. N. Majerus, "Industrial Melanism in the Peppered Moth, *Biston betularia*: An Excellent Teaching Example of Darwinian Evolution in Action," *Evolution: Education and Outreach* 2 (2009): 63–74.

7. Ibid.

8. Western Forestry Leadership Association, "The True Cost of Wildfire in the Western U.S.," http://www.wflccenter.org/news_pdf/324_pdf.pdf (accessed December 5, 2010).

9. F. P. Rowe and R. Redfern, "Toxicity Tests on Suspected Warfarin Resistant House Mice (*Mus musculus* L.)," *Journal of Hygiene* 63 (1965): 417–25; F. J. MacSwiney and M. E. Wallace, "Genetics of Warfarin-Resistance in House Mice of Three Separate Localities," *Journal of Hygiene* 80 (1978): 69–75.

10. A. Romero et al., "Insecticide Resistance in the Bed Bug: A Factor in the Pest's Sudden Resurgence?" *Journal of Medical Entomology* 44 (2007): 175–78; K. S. Yoon et al., "Biochemical and Molecular Analysis of Deltamethrin Resistance in the Common Bed Bug (Hemiptera: Cimicidae)," *Journal of Medical Entomology* 45 (2008): 1092–1101.

11. H. Ranson et al., "Pyrethroid Resistance in African Anopheline Mosquitoes: What Are the Implications for Malaria Control?" *Trends in Parasitology* 27 (2011): 91–98.

12. H. Tabata, "Paclitaxel Production by Plant-Cell-Culture Technology," *Advances in Biochemical Engineering/Biotechnology* 87 (2004): 1–23.

13. United States Environmental Protection Agency, "2004 Greener Synthetic Pathways Award," http://www.epa.gov/greenchemistry/pubs/pgcc/winners/gspa04.html (accessed December 31, 2010).

14. H. C. Stutz, M. R. Stutz, and S. C. Sanderson, "*Atriplex robusta* (Chenopodiaceae), a New Perennial Species from Northwestern Utah," *Madroño* 48 (2001): 112–15.

EPILOGUE

1. J. B. Johnson et al., "Evolution Education in Utah: A State Office of Education–University Partnership Focuses on Why Evolution Matters," *Evolution: Education and Outreach* 2 (2009): 349–58.

2. Gallup, "Evolution, Creationism, Intelligent Design," http://www.gallup.com/poll/21814/evolution-creationism-intelligent-design.aspx (accessed April 3, 2011); Pew Forum on Religion and Public Life, *U.S. Religious Landscape Survey: Religious Beliefs and Practices: Diverse and Politically Relevant* (Washington, DC: Pew Research Center, 2008).

3. Gallup, "Four in 10 Americans Believe in Strict Creationism," http://www.gallup.com/poll/145286/Four-Americans-Believe-Strict-Creationism.aspx (accessed March 26, 2011).

4. J. E. Jones III, *"Tammy Kitzmiller et al. v. Dover Area School District et al.*: Memorandum Opinion," http://www.pamd.uscourts.gov/kitzmiller/kitzmiller_342.pdf (accessed March 27, 2011).

5. S. C. Meyer, "Teach the Controversy," *Discovery Institute—Center for Science and Culture*, March 30, 2002, http://www.discovery.org/a/1134 (accessed March 27, 2011).

6. For more detailed discussions on the flaws in writings of the small number of scientists who dispute evolution, see chapter 6 of E. C. Scott, *Evolution vs. Creationism*, 2nd ed. (Berkeley: University of California Press, 2009); chapter 5 of K. R. Miller, *Finding Darwin's God* (New York: Harper Perennial, 2000); and chapter 9 of my own book: D. J. Fairbanks, *Relics of Eden: The Powerful Evidence of Evolution in Human DNA*, updated paperback ed. (Amherst, NY: Prometheus Books, 2010).

7. The following recent study at the University of Arizona revealed how seriously deficient scientific literacy is among students: H. Sugarman et al., "Astrology Beliefs among Undergraduate Students," *Astronomy Education Review* 10 (2011): 010101.

8. Gallup, "Four in 10 Americans Believe in Strict Creationism."

9. The most reliable poll on the views of scientists is one conducted in 2009, published as a pdf document by the Pew Research Center for the People and the Press, "Scientific Achievements Less Prominent Than a Decade Ago," July 9, 2009, http://people-press.org (accessed March 27, 2011). This poll showed that 87 percent of scientists belonging to the American Association for the Advancement of Science agree that "humans and other living things have evolved over time and that evolution is the result of natural processes such as natural selection" (p. 5). Interestingly, the proportion of scientists who accept evolution has become a hot topic among opponents of evolution, and also one for humor among scientists. The Discovery Institute in 2001 launched a petition for scientists to sign if they do not support evolution. In February 2007, the Discovery Institute announced that "over 700 scientists from around the world have now signed a statement expressing their skepticism about the contemporary theory of Darwinian evolution" (http://www.discovery.org/a/2732 [accessed March 27, 2011]). To demonstrate the silliness of such a petition, the National Center for Science

Education (NCSE) launched its own "tongue-in-cheek parody" petition called "Project Steve" for scientists to sign if they supported evolution, but it came with a catch. Only those with the name Steve (in honor of Stephen J. Gould), or a derivative of it, such as Stephanie, could sign the petition. Even though only about 1 percent of scientists have such a name, the NCSE bet that it could obtain more signatures than the Discovery Institute. As of February 24, 2011, the NCSE "Steve-O-Meter" was at 1,157 (http://ncse.com/taking-action/project-steve [accessed March 27, 2011]). The most recent announcement by the Discovery Institute in 2007 touted the list as being "over 700," although I downloaded the list, updated January 2010, and a quick count, aided by spreadsheet software, showed 819 signatures (http://www.dissentfrom darwin.org [accessed March 27, 2011]). Evidently, Project Steve is ahead, even with restrictions excluding 99 percent of scientists from participation.

10. Charles R. Darwin, *On the Origin of Species by Means of Natural Selection, or the Preservation of Favoured Races in the Struggle for Life*, 6th ed. (London: John Murray, 1872), p. 421.

BIBLIOGRAPHY

Aiello, L. C. "Five Years of *Homo floresiensis*." *American Journal of Physical Anthropology* 142 (2008): 167–79.

Apuane Geopark. "Inventory Card of the Geosites in the Apuan Alps Regional Park. No. 247: Fossils in Marble Banks in Foce di Pianza." http://www.parcapuane.toscana.it/apuane _geopark/ENGLISH%20VERSION/apuanegeopark_geosites.html (accessed December 16, 2010).

Arora, G., N. Polavarapu, and J. F. McDonald. "Did Natural Selection for Increased Cognitive Ability in Humans Lead to an Elevated Risk of Cancer?" *Medical Hypotheses* 73 (2009): 453–56.

Avery, O. T., C. M. MacLeod, and M. McCarty. "Studies on the Chemical Nature of the Substance Inducing Transformation of the Pneumococcal Types." *Journal of Experimental Medicine* 79 (1944): 137–58.

Balzotti, M. R. B., J. N. Thornton, P. J. Maughan, D. A. McClellan, M. R. Stevens, E. N. Jellen, D. J. Fairbanks, and C. E. Coleman. "Expression and Evolutionary Relationships of the *Chenopodium* Quinoa 11S Seed Storage Protein Gene." *International Journal of Plant Science* 169 (2008): 281–91.

Bar-Maor, J. A., K. M. Kesner, and J. K. Karftori. "Human Tails." *Journal of Bone and Joint Surgery* 62 (1980): 508–10.

Behe, M. H. *The Edge of Evolution: The Search for the Limits of Darwinism*. New York: Free Press, 2007.

Belyi, V. A., P. Ak, E. Markert, H. Wang, W. Hu, A. Puzio-Kuter, and A. J. Levine. "The Origins and Evolution of the p53 Family of Genes." *Cold Spring Harbor Perspectives in Biology* 2 (2010): a001198.

Berger, L. R., D. J. de Ruiter, S. E. Churchill, P. Schmid, K. J. Carlson, P. H. Dirks, and J. M. Kibii. "*Australopithecus sediba*: A New Species of *Homo*-Like Australopith from South Africa." *Science* 328 (2010): 195–204.

Bird, C. P., B. E. Stranger, M. Liu, D. J. Thomas, C. E. Ingle, C. Beazley, W. Miller, M. E. Hurles, and E. T. Dermitzakis. "Fast-Evolving Noncoding Sequences in the Human Genome." *Genome Biology* 8 (2007): R118.

Booth, H. A. F., and P. W. H. Holland. "Eleven Daughters of *NANOG*." *Genomics* 84 (2004): 229–38.

Borlaug Global Rust Initiative. http://www.globalrust.org (accessed April 12, 2011).

Britten, R. J. "Coding Sequences of Functioning Human Genes Derived Entirely from Mobile

Element Sequences." *Proceedings of the National Academy of Sciences, USA* 101 (2004): 16825–30.

Brouilette, S. W., J. S. Moore, A. D. McMahon, J. R. Thompson, I. Ford, J. Shepherd, C. J. Packard, N. J. Samani, and West of Scotland Coronary Prevention Study Group. "Telomere Length, Risk of Coronary Heart Disease, and Statin Treatment in the West of Scotland Primary Prevention Study: A Nested Case-Control Study." *Lancet* 369 (2007): 107–14.

Carrier, D. R. "The Evolution of Locomotor Stamina in Tetrapods: Circumventing a Mechanical Constraint." *Paleobiology* 13 (1987): 326–41.

Castañeda, I. S., S. Mulitza, E. Schefuss, R. A. Lopes dos Santos, J. S. Sinninghe Damsté, and S. Schouten. "Wet Phases in the Sahara/Sahel Region and Human Migration Patterns in North Africa." *Proceedings of the National Academy of Sciences, USA* 106 (2009): 20159–63.

Cavalli-Sforza, L. L., and F. Cavalli-Sforza. *The Great Human Diasporas: The History of Diversity and Evolution.* Reading, MA: Addison-Wesley, 1995.

Chimpanzee Sequencing and Analysis Consortium. "Initial Sequence of the Chimpanzee Genome and Comparison with the Human Genome." *Nature* 437 (2005): 69–87.

Chou, H. H., T. Hayakawa, S. Diaz, M. Krings, E. Indriati, M. Leakey, S. Pääbo, Y. Satta, N. Takahata, and A. Varki. "Inactivation of CMP-N-acetylneuraminic Acid Hydroxylase Occurred prior to Brain Expansion during Human Evolution." *Proceedings of the National Academy of Sciences, USA* 99 (2002): 11736–41.

Collins, F. S. *The Language of God: A Scientist Presents Evidence for Belief.* New York: Free Press, 2006.

Collodel, G., E. Moretti, S. Capitani, P. Piomboni, C. Anichini, M. Estenoz, and B. Baccetti. "TEM, FISH and Molecular Studies in Infertile Men with Pericentric Inversion of Chromosome 9," *Andrologia* 38 (2006): 122–27.

Conti, M. E., and G. Cecchettib. "Biological Monitoring: Lichens as Bioindicators of Air Pollution Assessment—A Review." *Environmental Pollution* 114 (2001): 471–92.

Cook-Deegan, R., C. DeRienzo, J. Carbone, S. Chandrasekharan, C. Heaney, and C. Conover. "Impact of Gene Patents and Licensing Practices on Access to Genetic Testing for Inherited Susceptibility to Cancer: Comparing Breast and Ovarian Cancers with Colon Cancers." *Genetics in Medicine* 12 (2010): S15–38.

Coop, G., K. Bullaughey, F. Luca, and M. Przeworski. "The Timing of Selection at the Human *FOXP2* Gene." *Molecular Biology and Evolution* 25 (2008): 1257–59.

Coyne, J. *Why Evolution Is True.* New York: Viking, 2009.

Crespi, B. J., and K. Summers. "Positive Selection in the Evolution of Cancer." *Biological Review of the Cambridge Philosophical Society* 81 (2006): 407–24.

Cunningham, C. B., N. Schilling, C. Anders, and D. R. Carrier. "The Influence of Foot Posture on the Cost of Transport in Humans." *Journal of Experimental Biology* 213 (March 1, 2010): 790–97.

Cyran, K. A., and M. Kimmel. "Alternatives to the Wright-Fisher Model: The Robustness of Mitochondrial Eve Dating," *Theoretical Population Biology* 78 (2010): 165–72.

Damond, F., M. Worobey, P. Campa, I. Farfara, G. Colin, S. Matheron, F. Brun-Vézinet, D. L. Robertson, and F. Simon. "Identification of a Highly Divergent HIV Type 2 and Proposal for a Change in HIV Type 2 Classification." *AIDS Research and Human Retroviruses* 20 (2004): 666–72.

Darwin, C. R. *Narrative of the Surveying Voyages of His Majesty's Ships* Adventure *and* Beagle *between the Years 1826 and 1836, Describing Their Examination of the Southern Shores of South America, and the* Beagle's *Circumnavigation of the Globe. Journal and Remarks. 1832–1836*, 3 vols. London: Henry Colburn, 1839.

———. *On the Origin of Species by Means of Natural Selection, or the Preservation of Favoured Races in the Struggle for Life*. London: John Murray, 1859.

———. *On the Origin of Species by Means of Natural Selection, or the Preservation of Favoured Races in the Struggle for Life*. 6th ed. London: John Murray, 1872.

Darwin, F., ed. *The Life and Letters of Charles Darwin*. 2 vols. New York: D. Appleton and Company, 1887.

Dávalos, I. P., F. Rivas, A. L. Ramos, C. Galaviz, L. Sandoval, and H. Rivera. "Inv(9)(p24q13) in Three Sterile Brothers." *Annals of Genetics* 43 (2000): 51–54.

Dawkins, R. *The Blind Watchmaker*. New York: W. W. Norton, 1986.

———. *The Greatest Show on Earth: The Evidence for Evolution*. New York: Free Press, 2009.

de Heinzelin, J., J. D. Clark, T. White, W. Hart, P. Renne, G. WoldeGabriel, Y. Beyene, and E. Vrba. "Environment and Behavior of 2.5-Million-Year-Old Bouri Hominids." *Science* 284 (1999): 625–29.

de Knijff, P. "Messages through Bottlenecks: On the Combined Use of Slow and Fast Evolving Polymorphic Markers on the Human Y Chromosome." *American Journal of Human Genetics* 67 (2000): 1055–61.

Derenko, M. V., T. Grzybowski, B. A. Malyarchuk, J. Czarny, D. Miścicka-Śliwka, and I. A. Zakharov. "The Presence of Mitochondrial Haplogroup X in Altaians from South Siberia." *American Journal of Human Genetics* 69 (2001): 237–41.

Dewannieux, M., F. Harper, A. Richaud, C. Letzelter, D. Ribet, G. Pierron, and T. Heidmann. "Identification of an Infectious Progenitor for the Multiple-Copy HERV-K Human Endogenous Retroelements." *Genome Research* 16 (2006): 1548–56.

Diamond, J. *Guns, Germs, and Steel*. New York: W. W. Norton, 1999.

Discovery Institute. "Ranks of Scientists Doubting Darwin's Theory on the Rise." Discovery Institute—Center for Science and Culture, February 8, 2007. http://www.discovery.org/a/2732 (accessed March 27, 2011).

Dobzhansky, T. "Nothing in Biology Makes Sense except in the Light of Evolution." *American Biology Teacher* 35 (1973): 125–29.

Dorus, S., E. J. Vallender, P. D. Evans, J. R. Anderson, S. L. Gilbert, M. Mahowald, G. J. Wyckoff, C. M. Malcolm, and B. T. Lahn. "Accelerated Evolution of Nervous System Genes in the Origin of *Homo sapiens*." *Cell* 119 (2004): 1027–40.

Drets, M. E., and M. W. Shaw. "Specific Banding Patterns of Human Chromosomes." *Proceedings of the National Academy of Sciences, USA* 68 (1971): 2073–77.

Duan, J. J., M. Marvier, J. Huesing, G. Dively, and Z. Y. Huang. "A Meta-Analysis of Effects of *Bt* Crops on Honey Bees (Hymenoptera: Apidae)." *PLoS ONE* 3 (2008): e1415.

Dubin, H. J., and J. P. Brennan. *Combating Stem and Leaf Rust of Wheat: Historical Perspective, Impacts, and Lessons Learned.* Washington, DC: International Food Policy Research Institute, 2009.

Dumont, P., J. I. Leu, A. C. Della Pietra III, D. L. George, and M. Murphy. "The Codon 72 Polymorphic Variants of p53 Have Markedly Different Apoptotic Potential." *Nature Genetics* 33 (2003): 357–65.

El Hallani, S., F. Ducray, A. Idbaih, Y. Marie, B. Boisselier, C. Colin, F. Laigle-Donadey, M. Rodéro, O. Chinot, J. Thillet, K. Hoang-Xuan, J. Y. Delattre, and M. Sanson. "*TP53* Codon 72 Polymorphism Is Associated with Age at Onset of Glioblastoma." *Neurology* 72 (2009): 332–36.

Ely, J. J., M. Leland, M. Martino, W. Swett, and C. M. Moore. "Technical Note: Chromosomal and mtDNA Analysis of Oliver." *American Journal of Physical Anthropology* 105 (1998): 395–403.

Etkind, A. "Beyond Eugenics: The Forgotten Scandal of Hybridizing Humans and Apes." *Studies in History and Philosophy of Biological and Biomedical Sciences* 39 (2008): 205–10.

Fairbanks, D. J. *Relics of Eden: The Powerful Evidence of Evolution in Human DNA.* Amherst, NY: Prometheus Books, 2007.

Fairbanks, D. J., and B. Rytting. "Mendelian Controversies: A Botanical and Historical Review." *American Journal of Botany* 88 (2001): 737–52.

Fairbanks, D. J., and P. J. Maughan. "Evolution of the *NANOG* Pseudogene Family in the Human and Chimpanzee Genomes." *BMC Evolutionary Biology* 6 (2006): 12.

Filchak, K. E., J. B. Roethele, and J. L. Feder. "Natural Selection and Sympatric Divergence in the Apple Maggot *Rhagoletis pomonella*." *Nature* 407 (2000): 739–42.

Finlayson, C., F. G. Pacheco, J. Rodríguez-Vidal, D. A. Fa, J. M. Gutierrez López, A. Santiago Pérez, G. Finlayson, E. Allue, J. Baena Preysler, I. Cáceres, J. S. Carrión, Y. Fernández Jalvo, C. P. Gleed-Owen, F. J. Jimenez Espejo, P. López, J. A. López Sáez, J. A. Riquelme Cantal, A. Sánchez Marco, F. G. Guzman, K. Brown, N. Fuentes, C. A. Valarino, A. Villalpando, C. B. Stringer, F. Martinez Ruiz, and T. Sakamoto. "Late Survival of Neanderthals at the Southernmost Extreme of Europe." *Nature* 443 (2006): 850–53.

Fleming, M. A., J. D. Potter, C. J. Ramirez, G. K. Ostrander, and E. A. Ostrander. "Understanding Missense Mutations in the *BRCA1* Gene: An Evolutionary Approach." *Proceedings of the National Academy of Sciences, USA* 100 (2003): 1151–56.

Franklin, A., A. W. F. Edwards, D. J. Fairbanks, D. L. Hartl, and T. Seidenfeld. *Ending the Mendel-Fisher Controversy.* Pittsburgh: University of Pittsburgh Press, 2007.

Frieden, T. R., T. Sterling, A. Pablos-Mendez, J. O. Kilburn, G. M. Cauthen, and S. W. Dooley.

"The Emergence of Drug-Resistant Tuberculosis in New York City." *New England Journal of Medicine* 328 (1993): 521–26.

Friedlaender, J., T. Schurr, F. Gentz, G. Koki, F. Friedlaender, G. Horvat, P. Babb, S. Cerchio, F. Kaestle, M. Schanfield, R. Deka, R. Yanagihara, and D. A. Merriwether. "Expanding Southwest Pacific Mitochondrial Haplogroups P and Q." *Molecular Biology and Evolution* 22 (2005): 1506–17.

Frisvold, G., R. Tronstad, and J. Mortensen. "Effects of *Bt* Cotton Adoption: Regional Differences in Producer Costs and Returns." Final Research Project Report Cotton Incorporated, January 1999, University of Arizona, Department of Agricultural and Resource Economics. http://www.bt.ucsd.edu/bt_cotton.html (accessed November 21, 2010).

Gallup. "Evolution, Creationism, Intelligent Design." http://www.gallup.com/poll/21814/evolution-creationism-intelligent-design.aspx (accessed April 3, 2011).

———. "Four in 10 Americans Believe in Strict Creationism." http://www.gallup.com/poll/145286/Four-Americans-Believe-Strict-Creationism.aspx (accessed March 26, 2011).

Garrison, K. E., R. B. Jones, D. A. Meiklejohn, N. Anwar, L. C. Ndhlovu, J. M. Chapman, A. L. Erickson, A. Agrawal, G. Spotts, F. M. Hecht, S. Rakoff-Nahoum, J. Lenz, M. A. Ostrowski, and D. F. Nixon. "T Cell Responses to Human Endogenous Retroviruses in HIV-1 Infection." *PLoS Pathology* 3 (2007): e165.

Genographic Project. http://www.genographic.org (accessed July 27, 2011).

Goidts, V., J. M. Szamalek, H. Hameister, and H. Kehrer-Sawatzki. "Segmental Duplication Associated with the Human-Specific Inversion of Chromosome 18: A Further Example of the Impact of Segmental Duplications on Karyotype and Genome Evolution in Primates." *Human Genetics* 115 (2004): 116–22.

Gonder, M. K., H. M. Mortensen, F. A. Reed, A. de Sousa, and S. A. Tishkoff. "Whole-mtDNA Genome Sequence Analysis of Ancient African Lineages." *Molecular Biology and Evolution* 24 (2007): 757–68.

Green, R. E., J. Krause, A. W. Briggs, T. Maricic, U. Stenzel, M. Kircher, N. Patterson, H. Li, W. Zhai, M. H. Fritz, N. F. Hansen, E. Y. Durand, A. S. Malaspinas, J. D. Jensen, T. Marques-Bonet, C. Alkan, K. Prüfer, M. Meyer, H. A. Burbano, J. M. Good, R. Schultz, A. Aximu-Petri, A. Butthof, B. Höber, B. Höffner, M. Siegemund, A. Weihmann, C. Nusbaum, E. S. Lander, C. Russ, N. Novod, J. Affourtit, M. Egholm, C. Verna, P. Rudan, D. Brajkovic, Z. Kucan, I. Gusic, V. B. Doronichev, L. V. Golovanova, C. Lalueza-Fox, M. de la Rasilla, J. Fortea, A. Rosas, R. W. Schmitz, P. L. Johnson, E. E. Eichler, D. Falush, E. Birney, J. C. Mullikin, M. Slatkin, R. Nielsen, J. Kelso, M. Lachmann, D. Reich, and S. Pääbo. "A Draft Sequence of the Neandertal Genome." *Science* 328 (2010): 710–22.

Greider, C. W., and E. H. Blackburn, "Identification of a Specific Telomere Terminal Transferase Activity in *Tetrahymena* Extracts. *Cell* 43 (1985): 405–13.

Greiner, T. M. "The Morphology of the Gluteus Maximus during Human Evolution: Prerequisite or Consequence of the Upright Bipedal Posture?" *Human Evolution* 17 (2006): 79–94.

Grochola, L. F., A. Vazquez, E. E. Bond, P. Würl, H. Taubert, T. H. Müller, A. J. Levine, G. L. Bond. "Recent Natural Selection Identifies a Genetic Variant in a Regulatory Subunit of Protein Phosphatase 2A that Associates with Altered Cancer Risk and Survival." *Clinical Cancer Research* 15 (2009): 6301–6308.

Haeckel, E. *Kunstformen der Natur*. Leipzig: Bibliographisches Institut, 1904.

Haile-Selassie, Y., G. Suwa, and T. D. White. "Late Miocene Teeth from Middle Awash, Ethiopia, and Early Hominid Dental Evolution." *Science* 303 (2004): 1503–1505.

Hammer, M. F., V. F. Chamberlain, V. F. Kearney, D. Stover, G. Zhang, T. Karafet, B. Walsh, and A. J. Redd. "Population Structure of Y Chromosome SNP Haplogroups in the United States and Forensic Implications for Constructing Y Chromosome STR Databases," *Forensic Science International* 164 (2006): 45–55.

Heap, I. M. "The Occurrence of Herbicide-Resistant Weeds Worldwide." *Pesticide Science* 51 (1997): 235–43.

Held, L. I., Jr. *Quirks of Human Anatomy: An Evo-Devo Look at the Human Body*. Cambridge: Cambridge University Press, 2009.

Hershey, A. D., and M. Chase. "Independent Functions of Viral Protein and Nucleic Acid in Growth of Bacteriophage." *Journal of General Physiology* 36 (1952): 39–56.

Hill, A. V., B. Gentile, J. M. Bonnardot, J. Roux, D. J. Weatherall, J. B. Clegg. "Polynesian Origins and Affinities: Globin Gene Variants in Eastern Polynesia." *American Journal of Human Genetics* 40 (1987): 453–63.

Hooper, J. *Of Moths and Men*. New York: W. W. Norton, 2002.

Huttley, G. A., S. Easteal, M. C. Southey, A. Tesoriero, G. G. Giles, M. R. McCredie, J. L. Hopper, and D. J. Venter. "Adaptive Evolution of the Tumour Suppressor *BRCA1* in Humans and Chimpanzees: Australian Breast Cancer Family Study." *Nature Genetics* 25 (2000): 410–13.

Inai, Y., Y. Ohta, and M. Nishikimi. "The Whole Structure of the Human Nonfunctional L-gulono-gamma-lactone Oxidase Gene—the Gene Responsible for Scurvy—and the Evolution of Repetitive Sequences Thereon." *Journal of Nutritional Science and Vitaminology* 49 (2003): 315–19.

Ingman, M., H. Kaessmann, S. Pääbo, and U. Gyllensten. "Mitochondrial Genome Variation and the Origin of Modern Humans." *Nature* 408 (2000): 708–13.

International Human Genome Sequencing Consortium. "Finishing the Euchromatic Sequence of the Human Genome." *Nature* 431 (2004): 931–45.

International Human Genome Sequencing Consortium. "Initial Sequencing and Analysis of the Human Genome." *Nature* 409 (2001): 860–921.

Jeter, C. R., M. Badeaux, G. Choy, D. Chandra, L. Patrawala, C. Liu, T. Calhoun-Davis, H. Zaehres, G. Q. Daley, and D. G. Tang. "Functional Evidence that the Self-Renewal Gene *NANOG* Regulates Human Tumor Development," *Stem Cells* 27 (2009): 993–1005.

Jin, Y., L. J. Szabo, Z. A. Pretorius, R. P. Singh, R. Ward, and T. Fetch, Jr. "Detection of Virulence to Resistance Gene *Sr24* within Race TTKS of *Puccinia graminis* f. sp. *tritici.*" *Plant Disease* 92 (2008): 923–26.

Johnson, J. B., M. Adair, B. J. Adams, D. J. Fairbanks, V. Itamura, D. E. Jeffery, D. Merrell, S. M. Ritter, and R. R. Tolman. "Evolution Education in Utah: A State Office of Education-University Partnership Focuses on Why Evolution Matters." *Evolution: Education and Outreach* 2 (2009): 349–58.

Jones, J. E., III, "*Tammy Kitzmiller et al. v. Dover Area School District et al.*: Memorandum Opinion." http://www.pamd.uscourts.gov/kitzmiller/kitzmiller_342.pdf (accessed March 27, 2011).

Jones, J. S., X. Chi, X. Gu, P. M. Lynch, C. I. Amos, and M. L. Frazier. "*p53* Polymorphism and Age of Onset of Hereditary Nonpolyposis Colorectal Cancer in a Caucasian Population." *Clinical Cancer Research* 10 (2004): 5845–49.

Jurka, J., and E. Zuckerkandl. "Free Left Arms as Precursor Molecules in the Evolution of *Alu* Sequences," *Journal of Molecular Evolution* 33 (1991): 49–56.

Karafet, T. M., F. L. Mendez, M. B. Meilerman, P. A. Underhill, S. L. Zegura, and M. F. Hammer. "New Binary Polymorphisms Reshape and Increase Resolution of the Human Y Chromosomal Haplogroup Tree." *Genome Research* 18 (2008): 830–38.

Kehrer-Sawatzki, H., J. M. Szamalek, S. Tanzer, M. Platzer, and H. Hameister. "Molecular Characterization of the Pericentric Inversion of Chimpanzee Chromosome 11 Homologous to Human Chromosome 9." *Genomics* 85 (2005): 542–50.

Klaatsch, H. *The Evolution and Progress of Mankind.* Translated by J. McCabe. New York: F. A. Stokes, 1923.

Knowles, D. G., and A. McLysaght. "Recent De Novo Origin of Human Protein-Coding Genes." *Genome Research* 19 (2009): 1752–59.

Krause, J., C. Lalueza-Fox, L. Orlando, W. Enard, R. E. Green, H. A. Burbano, J. J. Hublin, C. Hänni, J. Fortea, M. de la Rasilla, J. Bertranpetit, A. Rosas, and S. Pääbo. "The Derived *FOXP2* Variant of Modern Humans Was Shared with Neandertals," *Current Biology* 17 (2007): 1–5.

Krüger, S., A. Bier, C. Engel, E. Mangold, C. Pagenstecher, M. von Knebel Doeberitz, E. Holinski-Feder, G. Moeslein, K. Schulmann, J. Plaschke, J. Rüschoff, H. K. Schackert, and German Hereditary Non-Polyposis Colorectal Cancer Consortium. "The *p53* Codon 72 Variation Is Associated with the Age of Onset of Hereditary Non-Polyposis Colorectal Cancer (HNPCC)." *Journal of Medical Genetics* 42 (2005): 769–73.

Kurzrock, R., H. M. Kantarjian, B. J. Druker, M. Talpaz. "Philadelphia Chromosome-Positive Leukemias: From Basic Mechanisms to Molecular Therapeutics." *Annals of Internal Medicine* 138 (2003): 819–30.

Lai, C. S., S. E. Fisher, J. A. Hurst, F. Vargha-Khadem, and A. P. Monaco. "A Forkhead-Domain Gene Is Mutated in a Severe Speech and Language Disorder." *Nature* 413 (2001): 519–23.

Leakey, M. D., and R. L. Hay. "Pliocene Footprints in the Laetoli Beds at Laetoli, Northern Tanzania." *Nature* 278 (1979): 317–23.

Lee, Y. N., and P. D. Bieniasz. "Reconstitution of an Infectious Human Endogenous Retrovirus." *PLoS Pathology* 3 (2007): e10.

Liu, X. F., T. K. Bera, L. J. Liu, and I. Pastan. "A Primate-Specific POTE-Actin Fusion Protein Plays a Role in Apoptosis." *Apoptosis* 14 (2009): 1237–44.

Liu, Y. J., D. Zheng, S. Balasubramanian, N. Carriero, E. Khurana, R. Robilotto, and M. B. Gerstein. "Comprehensive Analysis of the Pseudogenes of Glycolytic Enzymes in Vertebrates: The Anomalously High Number of *GAPDH* Pseudogenes Highlights a Recent Burst of Retrotrans-Positional Activity." *BMC Genomics* 10 (2009): 480.

Loddenkemper, R., and B. Hauer. "Drug-Resistant Tuberculosis: A Worldwide Epidemic Poses a New Challenge." *Deutches Ärzteblatt International* 107 (2010): 10–19.

Lovejoy, C. O., B. Latimer, G. Suwa, B. Asfaw, and T. D. White. "Combining Prehension and Propulsion: The Foot of *Ardipithecus ramidus*." *Science* 326 (2009): 72, 72e1–72e8.

Lovejoy, C. O., G. Suwa, L. Spurlock, B. Asfaw, and T. D. White. "The Pelvis and Femur of *Ardipithecus ramidus*: The Emergence of Upright Walking." *Science* 326 (2009): 71, 71e1–71e6.

Lovejoy, C. O., S. W. Simpson, T. D. White, B. Asfaw, and G. Suwa. "Careful Climbing in the Miocene: The Forelimbs of *Ardipithecus ramidus* and Humans Are Primitive." *Science* 326 (2009): 70, 70e1–70e8.

MacSwiney, F. J., and M. E. Wallace. "Genetics of Warfarin-Resistance in House Mice of Three Separate Localities." *Journal of Hygiene* 80 (1978): 69–75.

Majerus, M. E. N. "Industrial Melanism in the Peppered Moth, *Biston betularia*: An Excellent Teaching Example of Darwinian Evolution in Action." *Evolution: Education and Outreach* 2 (2009): 63–74.

Malhi, R. S., and D. G. Smith. "Haplogroup X Confirmed in Prehistoric North America." *American Journal of Physical Anthropology* 119 (2002): 84–86.

Marin, M. C., C. A. Jost, L. A. Brooks, M. S. Irwin, J. O'Nions, J. A. Tidy, N. James, J. M. McGregor, C. A. Harwood, I. G. Yulug, K. H. Vousden, M. J. Allday, B. Gusterson, S. Ikawa, P. W. Hinds, T. Crook, and W. G. Kaelin Jr. "A Common Polymorphism Acts as an Intragenic Modifier of Mutant p53 Behaviour." *Nature Genetics* 25 (2000): 47–54.

Maughan, P. J., T. B. Turner, C. E. Coleman, D. B. Elzinga, E. N. Jellen, J. A. Morales, J. A. Udall, D. J. Fairbanks, and A. Bonifacio. "Characterization of Salt Overly Sensitive (*SOS1*) Gene Homoeologs in Quinoa (*Chenopodium quinoa* Willd)." *Genome* (2009): 647–57.

McBrearty, S., and N. G. Jablonski. "First Fossil Chimpanzee." *Nature* 437 (2005): 105–108.

McKusick, V. A. *Mendelian Inheritance in Man: A Catalog of Human Genes and Genetic Disorders.* 12th ed. Baltimore: Johns Hopkins University Press, 1998.

Mendel, G. "Experiments in Plant Hybridization." In W. Bateson, *Mendel's Principles of Heredity.* Translated by C. T. Druery. Cambridge: Cambridge University Press, 1913.

Meyer, S. C. "Teach the Controversy." Discovery Institute—Center for Science and Culture, March 30, 2002. http://www.discovery.org/a/1134 (accessed March 27, 2011).

Miller, K. W. *Finding Darwin's God: A Scientist's Search for Common Ground between God and Evolution*. New York: Cliff Street Books, 1999.

Montefalcone, G., S. Tempesta, M. Rocchi, and N. Archidiacono. "Centromere Repositioning." *Genome Research* 9 (1999): 1184–88.

Mwaengo, D. M., and F. J. Novembre. "Molecular Cloning and Characterization of Viruses Isolated from Chimpanzees with Pathogenic Human Immunodeficiency Virus Infections." *Journal of Virology* 72 (1998): 8976–87.

Nabhan, G. P. *Where Our Food Comes From: Retracing Nikolay Vavilov's Quest to End Famine*. Washington, DC: Island Press, 2009.

National Center for Science Education. "Project Steve." http://ncse.com/taking-action/project-steve (accessed March 27, 2011).

National Human Genome Research Institute. "Genetic Information Nondiscrimination Act (GINA) of 2008." http://www.genome.gov/24519851 (accessed July 26, 2011).

Njau, P. N., Y. Jin, J. Huerta-Espino, B. Keller, and R. P. Singh. "Identification and Evaluation of Sources of Resistance to Stem Rust Race Ug99 in Wheat." *Plant Disease* 94 (2010): 413–19.

Organisation for Economic Co-operation and Development. "OECD-FAO Agricultural Outlook 2010–2019." http://www.oecd.org/dataoecd/15/37/45599621.pdf (accessed December 31, 2010).

Ørsted, D. D. "Tumor Suppressor p53 Arg72Pro Polymorphism and Longevity, Cancer Survival, and Risk of Cancer in the General Population." *Journal of Experimental Medicine* 204 (2007): 1295–1301.

Owen, M. D. K., and I. A. Zelaya, "Herbicide-Resistant Crops and Weed Resistance to Herbicides." *Pest Management Science* 61 (2005): 301–11.

Pavlicek, A., V. N. Noskov, N. Kouprina, J. C. Barrett, J. Jurka, and V. Larionov. "Evolution of the Tumor Suppressor *BRCA1* in Primates: Implications for Cancer Predisposition." *Human Molecular Genetics* 13 (2004): 2737–51.

Pei, M. *The Story of Latin and the Romance Languages*. New York: Harper & Row, 1976.

Perego, U. A., A. Achilli, N. Angerhofer, M. Accetturo, M. Pala, A. Olivieri, H. H. Kashani, K. H. Ritchie, R. Scozzari, Q. P. Kong, N. M. Myres, A. Salas, O. Semino, H. J. Bandelt, S. R. Woodward, and A. Torroni. "Distinctive Paleo-Indian Migration Routes from Beringia Marked by Two Rare mtDNA Haplogroups." *Current Biology* 13 (2009): 1–8.

Pew Forum on Religion and Public Life. *U.S. Religious Landscape Survey: Religious Beliefs and Practices: Diverse and Politically Relevant*. Washington, DC: Pew Research Center, 2008.

Pew Research Center for the People and the Press. "Scientific Achievements Less Prominent Than a Decade Ago," July 9, 2009. http://people-press.org (accessed March 27, 2011).

Plantier, J. C., M. Leoz, J. E. Dickerson, F. de Oliveira, F. Cordonnier, V. Lemée, F. Damond, D. L. Robertson, and F. Simon. "A New Human Immunodeficiency Virus Derived from Gorillas." *Nature Medicine* 15 (2009): 871–72.

Powles, S. B. "Evolved Glyphosate-Resistant Weeds around the World: Lessons to be Learnt." *Pest Management Science* 64 (2008): 360–65.

Ptak, S. E., W. Enard, V. Wiebe, I. Hellmann, J. Krause, M. Lachmann, and S. Pääbo. "Linkage Disequilibrium Extends across Putative Selective Sites in *FOXP2*." *Molecular Biology and Evolution* 26 (2009): 2181–84.

Puente, X. S., G. Velasco, A. Gutiérrez-Fernández, J. Bertranpetit, M. C. King, and C. López-Otín. "Comparative Analysis of Cancer Genes in the Human and Chimpanzee Genomes." *BMC Genomics* 7 (2006): 15.

Quentin, Y. "Fusion of a Free Left *Alu* Monomer and a Free Right *Alu* Monomer at the Origin of the *Alu* Family in the Primate Genomes." *Nucleic Acids Research* 20 (1992): 487–93.

———. "Origin of the *Alu* Family: A Family of Alu-Like Monomers Gave Birth to the Left and the Right Arms of the *Alu* Elements." *Nucleic Acids Research* 20 (1992): 3397–3401.

Raichlen, D. A., A. D. Gordon, W. E. Harcourt-Smith, A. D. Foster, and W. R. Haas. "Laetoli Footprints Preserve Earliest Direct Evidence of Human-Like Bipedal Biomechanics." *PLoS ONE* 5, no. 3 (2010): e9769.

Ranson, H., R. N'guessan, J. Lines, N. Moiroux, Z. Nkuni, and V. Corbel. "Pyrethroid Resistance in African Anopheline Mosquitoes: What Are the Implications for Malaria Control?" *Trends in Parasitology* 27 (2011): 91–98.

Ray, J. *The Wisdom of God Manifested in the Works of Creation.* London: R. Harbin at the Prince's-Arms in St Paul's Church Yard, 1717.

Reich, D., R. E. Green, M. Kircher, J. Krause, N. Patterson, E. Y. Durand, B. Viola, A. W. Briggs, U. Stenzel, P. L. Johnson, T. Maricic, J. M. Good, T. Marques-Bonet, C. Alkan, Q. Fu, S. Mallick, H. Li, M. Meyer, E. E. Eichler, M. Stoneking, M. Richards, S. Talamo, M. V. Shunkov, A. P. Derevianko, J. J. Hublin, J. Kelso, M. Slatkin, and S. Pääbo. "Genetic History of an Archaic Hominin Group from Denisova Cave in Siberia." *Nature* 468 (2010): 1053–60.

Rhesus Macaque Genome Sequencing and Analysis Consortium. "Evolutionary and Biomedical Insights from the Rhesus Macaque Genome." *Science* 316 (2007): 222–34.

Rich, S. M., F. H. Leendertz, G. Xu, M. LeBreton, C. F. Djoko, M. N. Aminake, E. E. Takang, J. L. Diffo, B. L. Pike, B. M. Rosenthal, P. Formenty, C. Boesch, F. J. Ayala, and N. D. Wolfe. "The Origin of Malignant Malaria." *Proceedings of the National Academy of Sciences, USA* 106 (2009): 14902–14907.

Rodriguez-Delfin, L. A., V. E. Rubin-de-Celis, and M. A. Zago. "Genetic Diversity in an Andean Population from Peru and Regional Migration Patterns of Amerindians in South America: Data from Y Chromosome and Mitochondrial DNA." *Human Heredity* 51 (2001): 97–106.

Rogalla, P., B. Kazmierczak, A. M. Flohr, S. Hauke, J. Bullerdiek. "Back to the Roots of a New Exon—The Molecular Archaeology of a SP100 Splice Variant." *Genomics* 63 (2000): 117–22.

Romero, A., M. F. Potter, D. A. Potter, K. F. Haynes. "Insecticide Resistance in the Bed Bug: A Factor in the Pest's Sudden Resurgence?" *Journal of Medical Entomology* 44 (2007): 175–78.

Rook, L., L. Bondioli, M. Köhler, S. Moyà-Solà, and R. Macchiarelli. "Oreopithecus Was a Bipedal Ape after All: Evidence from the Iliac Cancellous Architecture." *Proceedings of the National Academy of Sciences, USA* 96 (1999): 8795–99.

Rowe, F. P., and R. Redfern. "Toxicity Tests on Suspected Warfarin Resistant House Mice (*Mus musculus* L.)." *Journal of Hygiene* 63 (1965): 417–25.

Rutkowski, R., K. Hofmann, and A. Gartner. "Phylogeny and Function of the Invertebrate p53 Superfamily." *Cold Spring Harbor Perspectives in Biology* 2 (2010): a001131.

Sanford, J. C. *Genetic Entropy and the Mystery of the Genome*. 3rd ed. Waterloo, NY: FMS, 2008.

Sanger, F., G. M. Air, B. G. Barrell, N. L. Brown, A. R. Coulson, J. C. Fiddes, C. A. Hutchison III, P. M. Slocombe, and M. Smith. "Nucleotide Sequence of Bacteriophage φX174 DNA." *Nature* 265 (1977): 687–95.

Sasagawa, I., M. Ishigooka, Y. Kubota, M. Tomaru, T. Hashimoto, and T. Nakada. "Pericentric Inversion of Chromosome 9 in Infertile Men." *International Urology and Nephrology* 30 (1998): 203–207.

Sawada, I., C. Willard, C. K. Shen, B. Chapman, A. C. Wilson, and C. W. Schmid. "Evolution of *Alu* Family Repeats since the Divergence of Human and Chimpanzee." *Journal of Molecular Evolution* 22 (1985): 316–22.

Sawyer, G. J., V. Deak, E. E. Sarmiento, R. Milner, and I. Tattersall. *The Last Human: A Guide to Twenty-Two Species of Extinct Humans*. New Haven, CT: Yale University Press, 2007.

Schuchat, A., B. P. Bell, and S. C. Redd. "The Science behind Preparing and Responding to Pandemic Influenza: The Lessons and Limits of Science." *Clinical Infectious Disease* 52 (2011): S8–S12.

Scientific Dissent from Darwinism. http://www.dissentfromdarwin.org (accessed March 27, 2011).

Scott, E. C. *Evolution vs. Creationism: An Introduction*. Westport, CT: Greenwood Press, 2004.

Semino, O., A. S. Santachiara-Benerecetti, F. Falaschi, L. L. Cavalli-Sforza, and P. A. Underhill. "Ethiopians and Khoisan Share the Deepest Clades of the Human Y-Chromosome Phylogeny." *American Journal of Human Genetics* 70 (2002): 265–68.

Sharp, P. M., and B. H. Hahn. "The Evolution of HIV-1 and the Origin of AIDS." *Philosophical Transactions of the Royal Society B* 365 (2010): 2487–94.

Shubin, N. *Your Inner Fish*. New York: Pantheon Books, 2008.

Silvestri, A. R., Jr., and I. Singh. "The Unresolved Problem of the Third Molar: Would People Be Better Off without It?" *Journal of the American Dental Association* 134:450–55.

Simpson, S. W., J. Quade, N. E. Levin, R. Butler, G. Dupont-Nivet, M. Everett, and S. Semaw. "A Female *Homo erectus* Pelvis from Gona, Ethiopia." *Science* 322 (2008): 1089–92.

Soares, P., J. A. Trejaut, J. H. Loo, C. Hill, M. Mormina, C. L. Lee, Y. M. Chen, G. Hudjashov, P. Forster, V. Macaulay, D. Bulbeck, S. Oppenheimer, M. Lin, and M. B. Richards. "Climate Change and Postglacial Human Dispersals in Southeast Asia." *Molecular Biology and Evolution* 25 (2008): 1209–18.

Soares, P., L. Ermini, N. Thomson, M. Mormina, T. Rito, A. Röhl, A. Salas, S. Oppenheimer, V. Macaulay, and M. B. Richards. "Correcting for Purifying Selection: An Improved Human Mitochondrial Molecular Clock." *American Journal of Human Genetics* 84 (2009): 740–59.

Srebniak, M., A. Wawrzkiewicz, A. Wiczkowski, W. Kaźmierczak, and A. Olejek. "Subfertile Couple with inv(2), inv(9) and 16qh+." *Journal of Applied Genetics* 45 (2004): 477–79.

Stanford, C. *Upright: The Evolutionary Key to Becoming Human.* Boston: Houghton Mifflin, 1980.

Stanyon, R., M. Rocchi, O. Capozzi, R. Roberto, D. Misceo, M. Ventura, M. F. Cardone, F. Bigoni, and N. Archidiacono. "Primate Chromosome Evolution: Ancestral Karyotypes, Marker Order and Neocentromeres." *Chromosome Research* 16 (2008): 17–39.

Stutz, H. C., M. R. Stutz, and S. C. Sanderson, "*Atriplex robusta* (Chenopodiaceae), A New Perennial Species from Northwestern Utah." *Madroño* 48 (2001): 112–15.

Sugarman, H., C. D. Impey, S. Buxner, and J. Antonellis. "Astrology Beliefs among Undergraduate Students." *Astronomy Education Review* 10 (2011): 0101018.

Susman, R. L. "Fossil Evidence for Early Hominid Tool Use." *Science* 265 (1994): 1570–73.

Suwa, G., B. Asfaw, R. T. Kono, D. Kubo, C. O. Lovejoy, and T. D. White. "The *Ardipithecus ramidus* Skull and Its Implications for Hominid Origins." *Science* 326 (2009): 68, 68e1–68e7.

Suwa, G., R. T. Kono, S. Katoh, B. Asfaw, and Y. Beyene. "A New Species of Great Ape from the Late Miocene Epoch in Ethiopia." *Nature* 448 (2007): 921–24.

Szamalek, J. M., V. Goidts, J. B. Searle, D. N. Cooper, H. Hameister, and H. Kehrer-Sawatzki. "The Chimpanzee-Specific Pericentric Inversions That Distinguish Humans and Chimpanzees Have Identical Breakpoints in *Pan troglodytes* and *Pan paniscus*." *Genomics* 87 (2006): 39–45.

Tabashnik, B. E., A. J. Gassmann, D. W. Crowder, and Y. Carriére. "Insect Resistance to *Bt* Crops: Evidence versus Theory." *Nature Biotechnology* 26 (2008): 199–202.

Tabata, H., "Paclitaxel Production by Plant-Cell-Culture Technology." *Advances in Biochemical Engineering/Biotechnology* 87 (2004): 1–23.

TalkOrigins Archive. "Fossil Hominids: The Evidence for Human Evolution." http://www.talkorigins.org/faqs/homs (accessed July 27, 2011).

Tattersall, I., and J. H. Schwartz. *Extinct Humans.* New York: Névraumont, 2000.

Torrents, D., M. Suyama, E. Zdobnov, and P. Bork. "A Genome-Wide Survey of Human Pseudogenes." *Genome Research* 13 (2003): 2559–67.

United Nations Population Division, Department of Economic and Social Affairs, United Nations Secretariat. *The World at Six Billion.* http://www.un.org/esa/population/publications/six billion/sixbillion.htm (accessed October 26, 2010).

United Nations Population Division. *World Population Prospects: The 2000 Revision.* New York: United Nations, 2000.

United States Department of Agriculture, Agricultural Research Service. "Ug99 an Emerging Virulent Stem Rust Race." http://www.ars.usda.gov/Main/docs.htm?docid=14649 (accessed April 12, 2011).

United States Department of Health and Human Services, Centers for Disease Control and Prevention. "Typhoid Fever." http://www.cdc.gov/nczved/divisions/dfbmd/diseases/typhoid_ fever (accessed July 27, 2011).

United States Environmental Protection Agency. "2004 Greener Synthetic Pathways Award." http://www.epa.gov/greenchemistry/pubs/pgcc/winners/gspa04.html (accessed December 31, 2010).

van Oven, N., and M. Kayser. "Updated Comprehensive Phylogenetic Tree of Global Human Mitochondrial DNA Variation." *Human Mutation* 30 (2009): E386–E394.

Vidal, N., M. Peeters, C. Mulanga-Kabeya, N. Nzilambi, D. Robertson, W. Ilunga, H. Sema, K. Tshimanga, B. Bongo, and E. Delaporte. "Unprecedented Degree of Human Immunodeficiency Virus Type 1 (HIV-1) Group M Genetic Diversity in the Democratic Republic of Congo Suggests that the HIV-1 Pandemic Originated in Central Africa." *Journal of Virology* 74 (2000): 10498–507.

Volodko, N. V., E. B. Starikovskaya, I. O. Mazunin, N. P. Eltsov, P. V. Naidenko, D. C. Wallace, and R. I. Sukernik. "Mitochondrial Genome Diversity in Arctic Siberians, with Particular Reference to the Evolutionary History of Beringia and Pleistocenic Peopling of the Americas." *American Journal of Human Genetics* 82 (2008): 1084–1100.

Ward, C. V., W. H. Kimbel, and D. C. Johanson, "Complete Fourth Metatarsal and Arches in the Foot of *Australopithecus afarensis.*" *Science* 331 (2011): 750–53.

Wells, J. *Icons of Evolution: Why Much of What We Teach about Evolution Is Wrong.* Washington, DC: Regnery, 2002.

Wells, S. *Deep Ancestry: Inside the Genographic Project.* Washington, DC: National Geographic Society, 2006.

Western Forestry Leadership Association. "The True Cost of Wildfire in the Western U.S." http://www.wflccenter.org/news_pdf/324_pdf.pdf (accessed December 5, 2010).

White, T. D., B. Asfaw, Y. Beyene, Y. Haile-Selassie, C. O. Lovejoy, G. Suwa, and G. WoldeGabriel. "*Ardipithecus ramidus* and the Paleobiology of Early Hominids." *Science* 326 (2009): 75–86.

World Health Organization. *Global Tuberculosis Control: Epidemiology, Strategy, Financing.* Geneva, Switzerland: WHO; WHO/HTM/TB/2009.411, 2009.

———. *World Malaria Report 2008.* Geneva, Switzerland: WHO Press, 2008.

Yoon, K. S., D. H. Kwon, J. P. Strycharz, C. S. Hollingsworth, S. H. Lee, and J. M. Clark. "Biochemical and Molecular Analysis of Deltamethrin Resistance in the Common Bed Bug (Hemiptera: Cimicidae)." *Journal of Medical Entomology* 45 (2008): 1092–1101.

Young, J., J. Ménétrey, and B. Goud. "*RAB6C* Is a Retrogene that Encodes a Centrosomal Protein Involved in Cell Cycle Progression." *Journal of Molecular Biology* 397 (2010): 69–88.

Yunis, J. J., J. R. Sawyer, and K. Dunham. "The Striking Resemblance of High Resolution G-Banded Chromosomes of Man and Chimpanzee." *Science* 208 (1980): 1145–48.

Zalmout, I. S., W. J. Sanders, L. M. Maclatchy, G. F. Gunnell, Y. A. Al-Mufarreh, M. A. Ali, A. A. Nasser, A. M. Al-Masari, S. A. Al-Sobhi, A. O. Nadhra, A. H. Matari, J. A. Wilson, and P. D. Gingerich. "New Oligocene Primate from Saudi Arabia and the Divergence of Apes and Old World Monkeys." *Nature* 466 (2010): 360–64.

Zegura, S. L., T. M. Karafet, L. A. Zhivotovsky, and M. F. Hammer. "High-Resolution SNPs and Microsatellite Haplotypes Point to a Single, Recent Entry of Native American Y Chromosomes into the Americas." *Molecular Biology and Evolution* 21 (2004): 164–75.

Zerjal, T., Y. Xue, G. Bertorelle, R. S. Wells, W. Bao, S. Zhu, R. Qamar, Q. Ayub, A. Mohyuddin, S. Fu, P. Li, N. Yuldasheva, R. Ruzibakiev, J. Xu, Q. Shu, R. Du, H. Yang, M. E. Hurles, E. Robinson, T. Gerelsaikhan, B. Dashnyam, S. Q. Mehdi, and C. Tyler-Smith. "The Genetic Legacy of the Mongols." *American Journal of Human Genetics* 72 (2003): 717–21.

Zhang, J., D. M. Webb, and O. Podlaha. "Accelerated Protein Evolution and Origins of Human-Specific Features: *FOXP2* as an Example." *Genetics* 162 (2002): 1825–35.

Zhang, J., X. Wang, M. Li, J. Han, B. Chen, B. Wang, J. Dai. "*NANOGP8* Is a Retrogene Expressed in Cancers," *FEBS Journal* 273 (2006): 1723–30.

Zhang, Z., P. M. Harrison, Y. Liu, and M. B. Gerstein. "Millions of Years of Evolution Preserved: A Comprehensive Catalog of the Processed Pseudogenes in the Human Genome." *Genome Research* 13 (2003): 2541–58.

Zimmer, C. *Smithsonian Intimate Guide to Human Origins.* Toronto, ON, Canada: Madison Press Books, Smithsonian Books, and Collins, 2005.

INDEX